Report of Investigations 9681

Demands on the Knee During Kneeling and Squatting Activities Common to Low-Seam Mining

Susan M. Moore, Ph.D., Jonisha P. Pollard, MS, William L. Porter, MS,
Sean Gallagher, Ph.D., Alan G. Mayton, CMSP, PE

DEPARTMENT OF HEALTH AND HUMAN SERVICES
Centers for Disease Control and Prevention
National Institute for Occupational Safety and Health
Office of Mine Safety and Health Research
Pittsburgh, PA • Spokane, WA

June 2011

This document is in the public domain and may be freely copied or reprinted.

Disclaimer

Mention of any company or product does not constitute endorsement by the National Institute for Occupational Safety and Health (NIOSH). In addition, citations to Web sites external to NIOSH do not constitute NIOSH endorsement of the sponsoring organizations or their programs or products. Furthermore, NIOSH is not responsible for the content of these Web sites.

The findings and conclusions in this report are those of the authors and do not necessarily represent the views of the National Institute for Occupational Safety and Health.

Ordering Information

To receive documents or other information about occupational safety and health topics, contact NIOSH at

>   Telephone: **1–800–CDC–INFO** (1–800–232–4636)
>   TTY: 1–888–232–6348
>   e-mail: cdcinfo@cdc.gov
>
>   or visit the NIOSH Web site at **www.cdc.gov/niosh**.

For a monthly update on news at NIOSH, subscribe to NIOSH *eNews* by visiting **www.cdc.gov/niosh/eNews**.

DHHS (NIOSH) Publication No. 2011–176

June 2011

SAFER • HEALTHIER • PEOPLE™

Contents

Executive Summary 1
Introduction 2
Methods 4
    Subjects 4
    Experimental Design 4
    Preliminary Measurements 6
    EMG Measurement 7
    Pressure Measurement 8
    Motion Analysis Measurement 10
    Additional Testing Equipment and Syncing 12
    Testing Procedure 12
    Data Analysis 13
Results 18
    EMG Data 18
    Pressure Data 21
    Net Forces and Moments and Resulting Joint Kinematics Data 28
Discussion of Results 33
    EMG 33
    Pressure 33
    Net Forces and Moments 34
    Implications of Results 34
Conclusions 37
Acknowledgments 38
References 39
Appendix A 42
Appendix B 47

Figures

Figure 1. Schematic showing each posture tested during the experiment. ... 5

Figure 2. Photographs of the articulated (A) and nonarticulated (B) kneepads used in this study. ... 6

Figure 3. Subject wearing pressure sensor in the preshaped 90° flexion position. ... 9

Figure 4. Pressure sensor layout with individual pressure sensing units identified during palpation as the patella (A) and the PTT (B) for a representative subject with the shaded cells identifying the pressure distribution while kneeling near full flexion and kneeling on one knee for the same subject. ... 9

Figure 5. A) Anatomical marker set and B) testing marker set of theupper and lower right side of the body. (R=Right, L=Left) ... 11

Figure 6. Diagram of external shank forces and reaction forces and moments for kneeling near full flexion with respect to the global coordinate system (GCS). ... 16

Figure 7. Diagram of external shank forces and reaction forces and moments for kneeling near full flexion with respect to the anatomical shank coordinate system (ASCS). ... 16

Figure 8. External force diagrams with respect to the anatomical shank coordinate system (ASCS). ... 17

Figure 9. Summary of EMG activity by posture and block position. Each posture and block position has two groups of bars representing the activity of the left and right thigh muscles. Note that the bars representing the left and right thigh muscles mirror their arrangement in the body if looked at from the superior aspect. ... 19

Figure 10. Interaction of kneepad condition and block position on EMG activity of the left biceps femoris across postures. ... 20

Figure 11. Mean pressure ratio at the PTT region across postures. (* indicates significant difference with $p < 0.05$) ... 21

Figure 12. Mean of the mean pressure at the patella region for the various postures. (*indicates significant difference with $p < 0.05$). ... 22

Figure 13. Mean of the mean pressure at the PTT region for the various postures. (*indicates significant difference with $p < 0.05$). ... 23

Figure 14. Mean of the maximum pressure at the patella region for the various postures. (*indicates significant difference with $p < 0.05$). ... 24

Figure 15. Mean of the maximum pressure at the PTT region for the various postures. (*indicates significant difference with $p < 0.05$). ... 25

Figure 16. Mean of the variance at the patella region for the various postures (*indicates significant difference with $p < 0.05$). ... 26

Figure 17. Mean of the variance at the PTT region for the various postures (*indicates significant difference with $p < 0.05$). ... 27

Figure 18. Typical pressure distribution (psi) for squatting (A) and kneeling near full flexion (B). ... 28

Figure 19. Knee angles (degrees) for all postures. Varus and internal rotation angles are positive. ... 29

Figure 20. Mean tibial forces normalized to body weight (% BW). Lateral, superior, and anterior forces are positive. 30

Figure 21. Mean tibial moments normalized to body weight times height (% BW*Ht). Extension, varus, and internal rotation moments are positive. 32

Figure 22: Pelvis coordinate system highlighting the location of the right hip joint center 43

Figure 23: Orientation of the ATCS and ASCS...44

Tables

Table 1. Summary of significant main effects and interactions for normalized activity of all muscles. 18

Acronyms and Abbreviations

AJC	Ankle joint center
ASCS	Anatomical shank coordinate system
ASIS	Anterior superior iliac spine
ATCS	Anatomical coordinate system of the thigh
L.ASIS	Left anterior superior iliac spine
R.ASIS	Right anterior superior iliac spine
HJC	Hip joint center
KJC	Knee joint center
MCS	Measured coordinate system
MTCS	Measured coordinate system of the thigh
MSCS	Measured coordinate system of the shank
BW	Body weight
Ht	Height
GCS	Global coordinate system
Ft/c	Thigh-calf contact force
Fh/g	Heel-gluteus contact force
F1	Forces at the foot
F2	Forces at the knee
WLL	Weight of the lower leg
One Knee	Kneeling on one knee
Near Full	Kneeling near full flexion
Near 90	Kneeling near 90° of knee flexion
Squat	Squatting
MSHA	Mine Safety and Health Administration
NIOSH	National Institute for Occupational Safety and Health
EMG	Electromyography
MVC	Maximum voluntary contraction
MAV	Mean amplitude value
PTT	Patellar tendon and tibial tubercle
ANOVA	Analysis of variance

Demands on the Knee During Kneeling and Squatting Activities Common to Low-Seam Mining

Susan M. Moore, Ph.D., Jonisha P. Pollard, MS, William L. Porter, MS,
Sean Gallagher, Ph.D., Alan G. Mayton, CMSP, PE

Office of Mine Safety and Health Research
National Institute for Occupational Safety and Health

Executive Summary

In 2009, the operating height of approximately one fourth of underground coal mines in the U.S. restricted mine workers to kneeling, crawling, and/or stooping posture to perform work [MSHA 2009]. The large number of knee injuries to these workers is likely attributed to exposure to musculoskeletal disorder risk factors (prolonged kneeling, crawling, and twisting on one's knees). Therefore, the National Institute for Occupational Safety and Health has investigated three different biomechanical parameters (muscle activity of the knee flexors and extensors, pressure at the knee, and the net forces and moments at the knee) as subjects assumed postures common to low-seam mining, both with and without kneepads. The postures evaluated included: (1) kneeling near full flexion; (2) kneeling near 90° of knee flexion; (3) kneeling on one knee; and (4) squatting. The pressure and the net forces and moments at the knee were evaluated as subjects statically assumed these postures. However, negligible muscle activity existed for these static postures. Therefore, muscle activity of the knee flexors and extensors was evaluated for each posture while subjects performed a lateral lift that is common to low-seam mining where they lifted a 25-lb block from their right side, transferred it across their body, and placed it on the ground on their left side. The results indicated that, relative to the stresses posed by other kneeling postures, some postures had may have more detrimental effects than others. Considering the potential impact of the three biomechanical parameters, several key recommendations were made regarding when it may be most appropriate to use specific postures. Additional recommendations were also made regarding the design of kneepads.

Introduction

Underground coal mining creates a unique set of environmental stressors not found in other occupations. These stressors include reduced visibility, uneven terrain, falling roof material, and other hazards which require mine workers to don numerous forms of personal protective equipment including hard hats, steel-toed boots, and cap lamps. However, one of the most challenging stressors is often the extreme restriction in vertical working height present in many underground coal mines. Although dependent on mining conditions, the working height of a mine is typically similar to the height of the coal seam. Seam height can be classified as low (≤ 43in) and medium (44 to 60 in) with higher seams being greater than 60 in. Low-seam heights restrict mine workers' postures, forcing them to kneel, crawl, and/or stoop to perform work. Mine workers in low-seam mines spend their entire shift (8–10 hours) confined to these postures. A previous NIOSH evaluation of the 2007 Mine Safety and Health Administration (MSHA) injury database indicated that 84 injuries to the knee were reported for seam heights of 30 to 54 in. These injuries were associated with an average of 47 days away from work for the low seams and 62 days for the mid seams. In an analysis of musculoskeletal injury data from eight low-seam coal mines, Gallagher et al. [2009] reported that the cost per knee injury was $13,121 on average. Considering the 84 injuries that occurred in mines with a seam height of 30 to 54 in, it can be estimated that these injuries cost more than $1,000,000 in 2007. The majority of these workers are more senior and male. According to a recent national survey of the mining industry conducted by NIOSH, the average age of coal mine workers (surface and underground mines combined) is 43.8 years (95% CI: 42.5 to 45.1 years). NIOSH also determined that, within this population, only 3.8% are female.

Due to the fact that mine workers are confined to their knees for the duration of their shift, many studies regarding knee injuries in mining have focused on this population of workers. These studies have demonstrated that low-seam mine workers suffer from a variety of knee injuries, including meniscal tears, osteoarthritis, ligament tears, and bursitis (i.e., beat knee) [McMillan and Nichols 2005, Roantree 1957, Sharrard 1963, Sharrard 1964, Sharrard and Liddell 1962, Watkins et al. 1958]. These injuries are likely attributable to low working heights, which confine workers to kneeling and squatting postures, both of which have been associated with knee injuries [Baker et al. 2002, Baker et al. 2003, Coggon et al. 2000, Cooper et al. 1994, Felson et al. 1991, Sharrard and Liddell 1962, Tanaka et al. 1985].

Although not mandated by the government, many low-seam coal mine workers choose to wear kneepads to protect their knees from the uneven mine floor which may contain sharp, jagged rocks. The design of these kneepads varies between articulated and nonarticulated styles; however, they typically employ an outer shell to protect from the surface conditions and an inner padding to provide cushioning to the knee. The types of materials used to form the outer shell and inner padding are quite variable. Even though it is clear that these kneepads protect the knees from cuts and scrapes, their effect on the muscle activity of the thigh and the forces, pressures, and moments at the knee while in postures associated with low-seam mining, was previously unknown. The high number of injuries to the knee in low-seam mining suggests that these kneepads are only minimally successful in diminishing the negative effects of these parameters.

The effect that postures assumed by low-seam mine workers and kneepads have on the knee must be investigated to improve the design of interventions such as postural rotation strategies, equipment/work station design, and kneepads. Muscle activity, pressure applied to the knee, and the net forces and moments at the knee are three biomechanical parameters that researchers have previously used to gain insight into the risk of knee injury [Ayoub et al. 1985a, Ayoub et al. 1985b, Gallagher and Hamrick 1992, Gibbons 1989, Nagura et al. 2002, Perry et al. 1975, Sharrard 1964]. In postures associated with low-seam mining, the demands on the flexor and extensor muscles of the thigh are expected to be different than that of a normal standing or seated work posture. These flexors and extensors may have to generate force to support activities such as lifting, and may also be called upon to stabilize the body in what may be relatively unstable postures. As such, some postures may result in an earlier onset of fatigue. The pressure at the knee has long been considered to be a key parameter leading to the onset of knee injuries. In the early 1960s, Sharrard [1964] investigated the effect of a shoveling task on the force and pressure at the knee and noted that the greatest portion of mine workers suffering from pre-patellar bursitis were those that assumed static kneeling postures. Previous research regarding knee forces and moments when kneeling and squatting has shown an increase in these parameters compared to walking. Perry et al. [1975] found significantly increased forces on the joint surfaces with flexion of greater than 30°. At 30° position, the required force was 210% body weight and at 60° the required force was 410% body weight. Nagura et al. [2002] determined the forces during squatting activities and found posterior forces increased by 50% over walking and stair climbing.

This study investigated three biomechanical parameters associated with knee loading that are potentially related to knee injury risk. In order to determine the effects that commonly used postures in low-seam mining and wearing or not wearing kneepads have on these parameters, three objectives were met for three kneepad conditions (no kneepad, articulated kneepad, and nonarticulated kneepad, both of which were reported by distributors to be very commonly used):

1. Examine the electromyographic (EMG) responses of knee flexors and extensors during a lateral lifting task in kneeling and squatting postures associated with low-seam mining (Note: Static trials were excluded from this analysis because only minimal activity was observed in these trials). It was hypothesized that changes in posture would result in altered muscular demands required to accomplish the lifting task, and would be reflected by changes in the magnitudes and patterns of EMG activity of the knee flexors and extensors.

2. For static postures associated with low-seam mining, determine the pressures applied to the landmarks of the knee identified during pilot testing as being responsible for transmitting the majority of load to the knee (patella, patellar tendon, and tibial tubercle). It was hypothesized that the pressure and pressure distribution at the knee would be significantly affected by wearing a kneepad and by the simulated posture. It was further hypothesized that the type of kneepad worn would significantly affect the pressure and pressure distribution at the knee.

3. For postures associated with low-seam mining, investigate the net externally applied forces and moments at the knee and resulting joint kinematics. It was hypothesized that significant differences will be detected in the loading profiles between kneeling and squatting, as well as between the low-flexion (kneeling on one knee and kneeling near 90° flexion) and higher-flexion (squatting and kneeling near full flexion) postures.

Methods

Subjects

Ten subjects (seven male, three female) participated in this study. The average age was 34±17 (mean±SD) years with an age range of 19 to 60 years. The average weight was 154±22 lbs and the average height was 66±3 in. Prior to participation in the study, each subject was asked a series of questions to determine if they had a history of serious injury to the knee. None of the subjects reported knee surgery. One subject was diagnosed with bursitis which did not require any intervention and a second subject had slight nerve damage due to a motorcycle accident. These subjects' data compared well with their eight counterparts and were, therefore, not excluded from the study. None of the subjects participating in this study were currently working in the low-seam mining industry or in occupations that required kneeling, crawling, or squatting. Prior to participating in the study, each subject read and signed an informed consent form which was approved by the NIOSH Human Subjects Review Board.

Experimental Design

In this study, the impact various postures had on the activity of the thigh muscles, pressure applied to the knee, and the net forces and moments at the knee were investigated. NIOSH researchers interviewed 48 low-seam mine workers, whereby the mine workers identified the postures they used to perform various mine tasks (e.g., building stoppings, hanging curtain). The postures examined in this study are those indicated by these mine workers and are reflected in Figure 1. Two working heights (38 in and 48 in) were investigated in this study. Low-seam mines are those with working heights of 43 in or less; therefore, the working heights in this study provided a typical working height classified as a low-seam height and a working height that was higher than a low-seam height but still required the mine worker to adopt kneeling and squatting postures to perform their tasks. The 10 in total difference of the two selected working heights allowed researchers to determine if differences in seam heights in and around the low-seam level would significantly affect muscle activity, pressure at the knee, and the net forces and moments at the knee.

Not all postures were performed in both seam heights; only postures that were reasonable for each seam height were investigated. For example, a mine worker would not kneel near 90° of knee flexion in a 38 in mine as this would be extremely difficult at such a low working height.

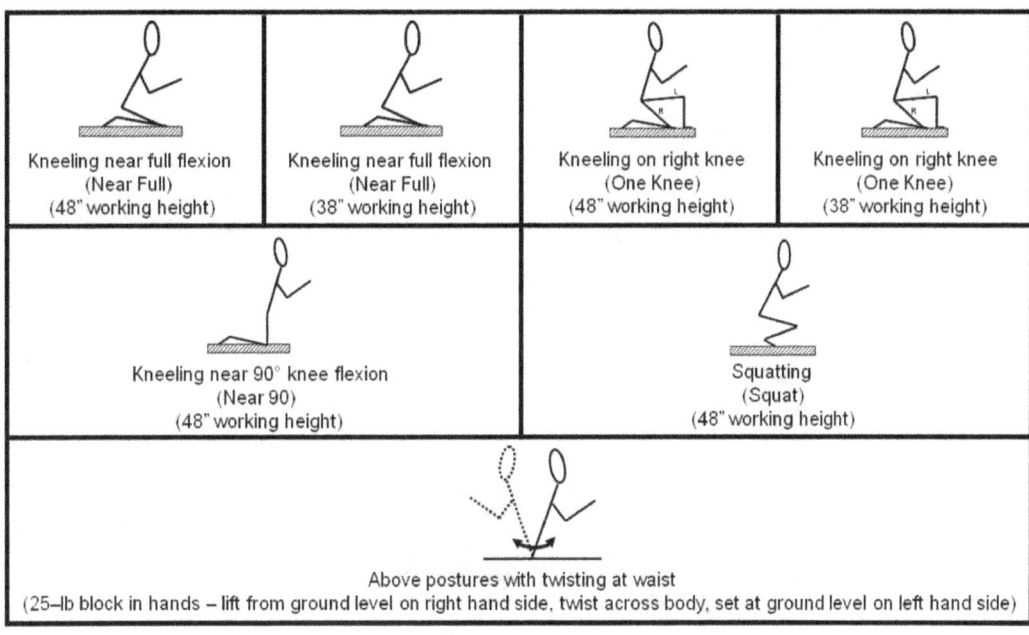

Figure 1. Schematic showing each posture tested during the experiment.

In addition to investigating the purely static postures depicted in the top two rows of Figure 1, several dynamic postures were also evaluated. It is quite common for mine workers to lift an object (e.g., stopping blocks, roof bolts) across their body while performing their daily duties; therefore, each of the static postures was also performed with an added dynamic component. The subjects would reach to their right to retrieve a 25-lb block, bring it in front of their body, and then set it down on their left side in one smooth motion. The block weight was selected as a representative weight considering the weight of stopping blocks (16–55 lbs) and the fact that most roof bolter operators move multiple roof bolts at a time (~7 lbs per roof bolt and varies by length). In this study, the 25-lb block that was used was positioned 9 in to the right of the lateral-most aspect of the right knee with the midline of the block in line with the front of the right knee. A target location where the subjects would set the block was also marked; this location was 9 in to the left of the lateral-most aspect of the left knee and marked such that the midline of the block, when set down, would be in line with the front of the left knee. This 9-in location was selected because it was comfortably within reach for individuals of many different anthropometries and did not interfere with other experimental equipment (e.g., force plates).

Subjects performed each posture for three kneepad conditions: no kneepads, articulated kneepads, and nonarticulated kneepads. The articulated and nonarticulated kneepads were selected due to popularity in the industry. Several distributors of kneepads to the mining industry were contacted in 2007 and asked to provide the most frequently ordered kneepads for the previous year. From these data, the most commonly requested articulated and nonarticulated kneepads were selected (Figure 2). The articulated kneepads consisted of a hard outer shell with hard rubber padding on the inside. The straps were also rubber and crossed a few inches above and a few inches below the crease of the knee. The nonarticulated kneepads consisted of a soft outer rubber shell and soft inner foam padding. Again, the straps crossed a few inches above and below the crease of the knee.

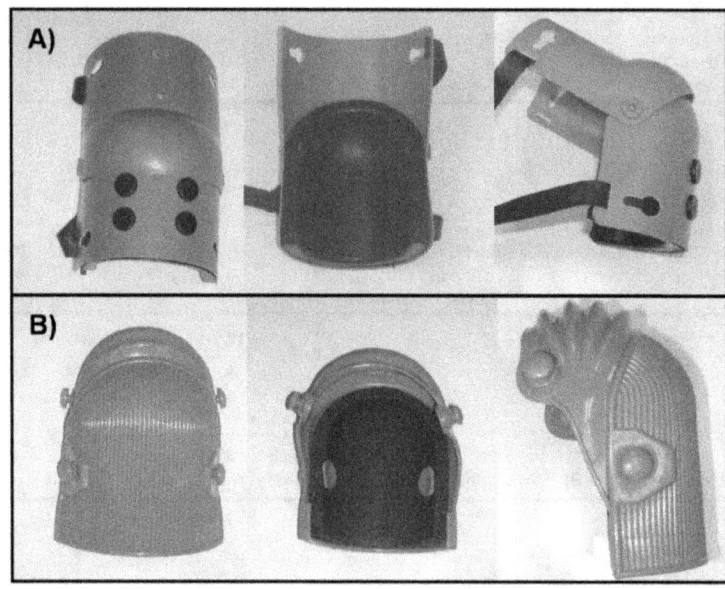

Figure 2. Photographs of the articulated (A) and nonarticulated (B) kneepads used in this study.

Preliminary Measurements

Preliminary measurements were taken prior to subject testing, including subject height, weight, and thigh-calf and heel-gluteus contact pressures. The contact forces were calculated from the contact pressure data. The locations of these forces were then determined with respect to the knee. Both the force magnitude and location of the force were necessary inputs to the inverse dynamics model described later in this study. Due to the complexity of these measurements, measuring the contact pressures during laboratory testing was not possible. Pilot testing showed standard repeatability error within 5% body weight for repeated measures of contact forces with and without kneepads. Therefore, these contact force measurements were collected prior to the start of any other data collection with the subject. These values were obtained using a clinical seating pressure sensor system that employs resistive technology (Tekscan ClinSeat®, South Boston, MA). For the thigh-calf contact force, the sensor was placed on the subject's lower right leg (from popliteal to the heel), and the subject was instructed to squat. The distance from the uppermost sensing unit to the lateral epicondyle of the right femur was then measured with a ruler, and pressure data was then collected for five seconds. The subject was then instructed to stand and relax. The sensor was then placed on the subject's lower right leg, crossing both the thigh and heel, and the subject was instructed to kneel near full flexion. In each case, the distance from the uppermost sensing unit to the lateral epicondyle of the right femur was measured. Data was again collected for five seconds. Using the Tekscan software, the area that represented the thigh-calf or heel-gluteus contact area was then isolated, and the total force and position of the center of pressure were reported with respect to the uppermost sensing unit. The average force magnitude and location of the center of pressure were then calculated.

EMG Measurement

All EMG measurements were made using a Noraxon Telemyo™ 2400R World Wide Telemetry system (Noraxon USA Inc., Scottsdale, AZ) with 16 channels. The gain was set to 10 for all channels. Several hardware filters were in place: 1st order high-pass filters set to 10 Hz ±10% cutoff, 8th order Butterworth/Bessels low-pass, anti-alias filters set to 500 Hz or 1,000 Hz±2% cutoff. The common mode rejection was > 100 dB and the sampling rate for electromyography data was set at 1020 Hz.

The longer of the two kneepad types, the articulated kneepads, were placed on the right and left knees of the subject; the superior-most aspect of the kneepads were marked using an ink pen. The location of the placement for each EMG electrode (disposable, self-adhesive Ag/AgCl dual snap surface electrodes with 2 cm spacing) was then measured and marked again using the ink pen. Prior to placement of the electrodes, any hair at the placement sites was shaved and the sites were then cleaned and abraded using an electrode skin prep pad. The muscles selected for EMG analysis were the right and left: rectus femoris, vastus medialis, vastus lateralis, semitendinosus, and biceps femoris. When possible, the electrodes were placed over the belly of the muscles, distal to the motor point regions [Ericson 1985]. In some cases, it was not possible to place the electrodes in the ideal location due to the length of the articulated kneepad. In those cases, the electrode was placed as closely as possible to its ideal location. For the rectus femoris, the distance from the anterior superior iliac spine (ASIS) to the apex of the patella was measured and the electrodes were placed at the midpoint of this distance. The electrodes were placed along the line running from the ASIS to the patellar apex. For the vastus medialis, the distance from the ASIS to the medial knee joint space was measured and the electrodes were placed at a point which was 80% of this distance. Electrodes were angled at approximately 20° from the midline of the body such that they ran along the length of the muscle belly. For the vastus lateralis, the distance from the ASIS to the lateral knee joint space was measured and the electrodes were placed at a point which was 75% of this distance. Electrodes were angled at approximately 15° from the midline of the body such that they ran along the length of the muscle belly. For the semitendinosus, the distance between the ischial tuberosity and the medial joint space was measured and the electrodes were placed slightly lateral to the midpoint of this distance along the long axis of the thigh. Lastly, for the biceps femoris, the distance between the ischial tuberosity and the lateral knee joint space was measured and electrodes were placed at the midpoint of this distance along the long axis of the thigh. A reference electrode (disposable, self-adhesive Ag/AgCl snap surface electrode) was also placed at a remote site above the greater trochanter.

Maximum voluntary contractions (MVCs) were then determined for the right and left thigh muscles. The subject was instructed to lay in a supine position in a Biodex™ chair (Biodex Medical Systems, Inc., Shirley, NY) with hip and knee angles at 90°. The subject was then instructed to extend his/her knee with maximal effort for at least 5 seconds while a researcher provided verbal encouragement. The subject was allowed to rest and was then instructed to flex the knee with maximal effort for at least 5 seconds again while verbally encouraged by a researcher.

Pressure Measurement

A custom-built pressure sensor (TactArray T2000; Pressure Profile Systems, Los Angeles, CA) was used to measure pressures at the knee. This sensor used capacitive sensor technology and was preshaped to conform to the knee when flexed at 90°. The sensor was 0.13-in thick and consisted of 196 individual pressure-sensing units that varied in size ranging from 0.17 in^2 to 0.23 in^2. Due to the preshaped nature of the sensor, the distance between the sensing units (dead space) was not constant across the sensor. With the sensor affixed to the knee, the distance between sensing units in the medial-to-lateral direction was constant at 0.07 in. However, in the superior-to-inferior direction, the distance between sensing units varied from 0.12 to 0.19 in within the region where the patellar tendon and tibial tubercle (PTT) rested and from 0.12 to 0.32 in within the region where the patella rested. Additionally, there was Velcro® sewn in to extra fabric on either end of the sensor to aide with fixation. To affix the sensor to the leg, the knee was held at 90° of flexion and an Ace® bandage was first wrapped around the thigh so that half of the bandage was unwrapped and the bottom edge of the bandage was 2 in from the crease of the knee. A rectangular piece of Velcro was then adhered such that half attached to the skin and the other half to the bandage. Eight pieces of hypoallergenic athletic tape were then applied in an asterisk pattern such that half of the tape was adhered to the Velcro on the thigh and the other half was adhered to the skin of the thigh. The corresponding piece of Velcro on the sensor was then adhered. The remainder of the Ace bandage was then wrapped around the thigh and metal fasteners were used to attach it to itself. In a similar manner, the inferior aspect of the sensor was adhered to the lower leg. Once the sensor was affixed to the subject's leg, a sheer stocking was placed on the leg covering the sensor. This helped protect the sensor from tears while it was dragged along the laboratory floor when the subject crawled into the posture of interest. An iterative and meticulous testing process was used to determine final details of the fixation method, wherein the sensor did not move while the subject performed the activities of interest.

The next task was to identify which sensing units corresponded to various anatomic landmarks (patella, patellar tendon, and tibial tubercle). Using a wooden dowel, a researcher palpated the perimeter of these landmarks; the sensing units corresponding to each landmark were identified (Figures 3 and 4). The same researcher palpated the anatomic landmarks for all subjects. Preliminary tests were conducted to determine the researcher's repeatability for palpating the landmarks. Based on this information, it was decided that the PTT would be grouped together because their small size resulted in unacceptable repeatability when palpated individually. Palpating the patella and the PTT were both repeatable to within 6.7% of their respective total areas. These anatomic landmarks were also palpated at the conclusion of testing to ensure that the sensor had not moved. The sampling rate for pressure data was approximately 5 Hz.

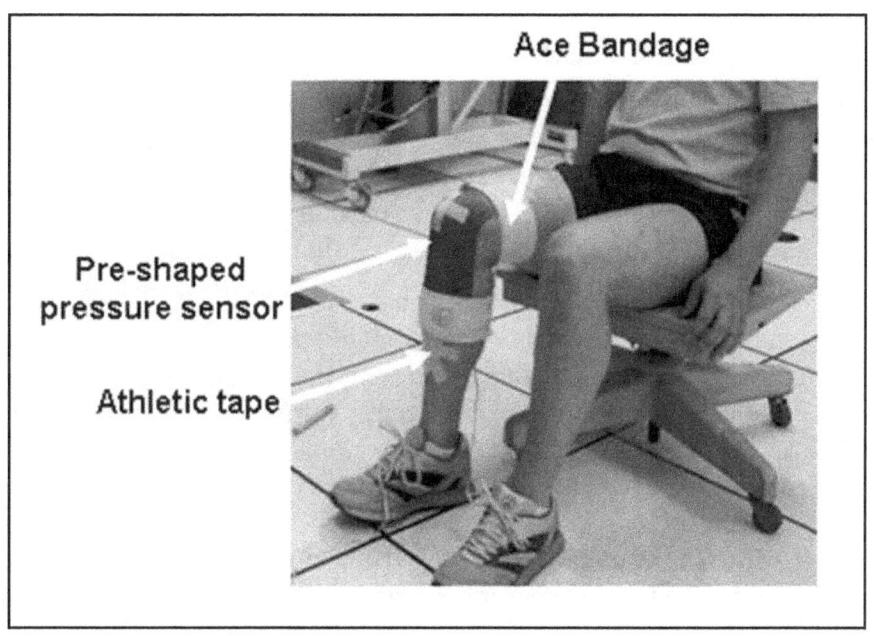

Figure 3. Subject wearing pressure sensor in the preshaped 90° flexion position.

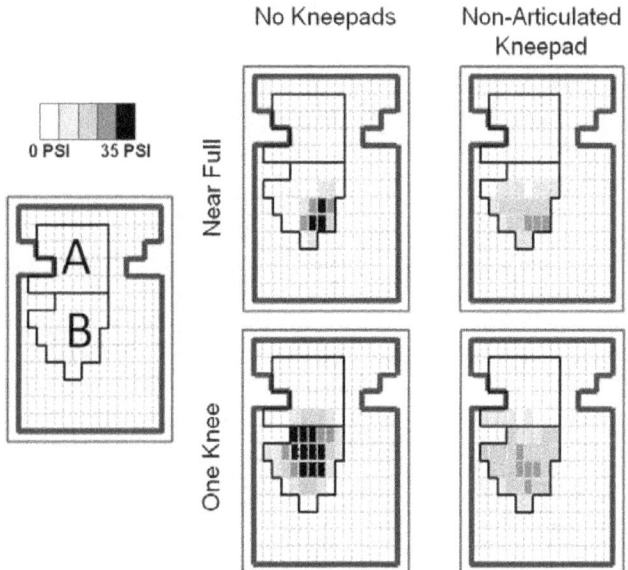

Figure 4. Pressure sensor layout with individual pressure sensing units identified during palpation as the patella (A) and the PTT (B) for a representative subject with the shaded cells identifying the pressure distribution while kneeling near full flexion and kneeling on one knee for the same subject.

Motion Analysis Measurement

An important part of this study was to quantitatively determine the position of the body and synchronize this data with a variety of other measurements (e.g., muscle activity, pressure at the knee, external forces). Therefore, a motion capture system (Eagle Digital RealTime System; Motion Analysis Corporation, Santa Rosa, CA) was used to quantitatively determine the location of all the body segments as the subject performed the various postures.

It was first necessary to establish a marker set that was ideal for the needs of this particular experiment, making it possible to determine joint center locations throughout testing. One unique problem with this study was that, with the application of the pressure sensor and kneepads, it was not possible to track the location of the knee joint center during the experiment. Instead, an initial anatomical marker set was developed that included markers on the medial and lateral aspects of the knee; a second marker set, where several markers were removed from the subject, was used during testing. The anatomical marker set included 41 markers; 8 markers were removed for the testing marker set (n=33). The markers comprising the anatomical marker set and testing marker set are shown in Figure 5. The testing marker set is a modified version of the Cleveland Clinic marker set [Kirtley 2006].

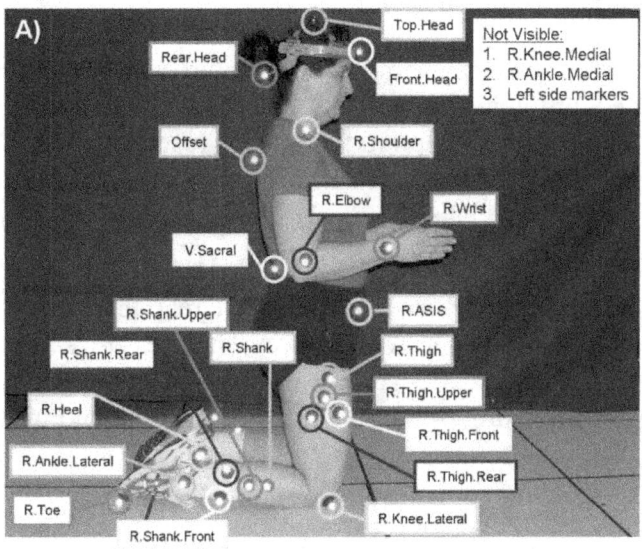
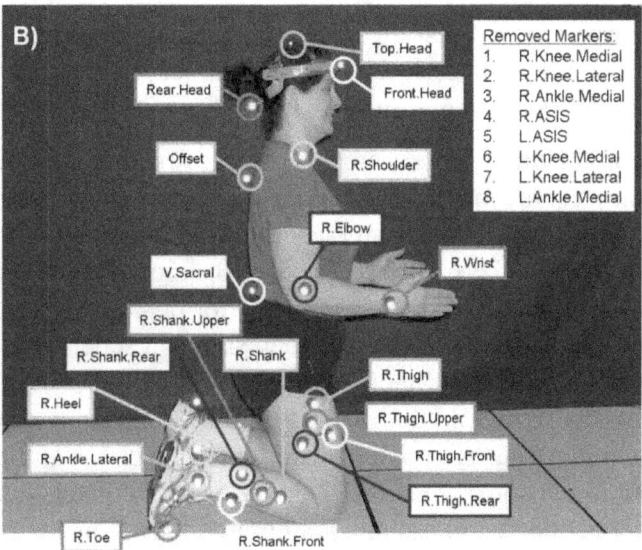

Figure 5. A) Anatomical marker set and B) testing marker set of the upper and lower right side of the body. (R=Right, L=Left)

Instrumented with the anatomical marker set, the subject was instructed to stand with their arms out to either side of their body making a "T". This is known as a "T-pose" and is used to determine the position of the anatomic landmarks (medial and lateral femoral epicondyles, medial malleolus, and right and left anterior superior iliac spine) with respect to the markers that were used during testing. After the T-pose data were collected, the eight markers were removed yielding the testing marker set. The sampling rate for motion capture data was 60 Hz.

Additional Testing Equipment and Syncing

In addition to the pressure sensor system, motion capture system, and EMG system, two force plates were also used during testing (OR6-5, Advanced Mechanical Technology, Inc., Watertown, MA). These force plates captured the ground-reaction forces applied to the foot and knee which would later be used as inputs to a computational model that solved for the net forces and moments at the knee (described later). Only the right knee was investigated in this study.

The force plate and EMG data were collected (1020 Hz) using the EvaRT 5.0.4 software (Motion Analysis Corporation, Santa Rosa, CA) and were read through an analog-to-digital board (PCI-6071E, National Instruments, Austin, TX). The pressure sensor system produced a sync signal that was output once data collection was initiated. This signal was also read through the analog-to-digital board; therefore, pressure, force plate, EMG, and motion capture data were all synched.

Testing Procedure

A randomized order was used for testing the three kneepad conditions (no kneepads, articulated kneepads, and nonarticulated kneepads). Within each kneepad condition, the order in which postures were tested was also randomized.

If the testing condition included a kneepad, the subject first donned the required kneepad. Then, for all tests, the subject was instructed to place their knee in a reference position in order to set the pressure sensor at zero. In this position the subject was seated with his/her knee at 90° for postures with knee angles near 90° (i.e., kneeling near 90° knee flexion and kneeling on one knee; see Figure 1). This reference position was a squat for postures with knees in high flexion (e.g., kneeling near full flexion). The subject was then shown a schematic (similar to Figure 1) of the posture they were to assume. The subjects were also instructed to keep their hands central to their body. Some subjects chose to hold their hands in the air around their chest while others let their hands rest on their thighs or at their sides. In some instances the subject inquired as to the whether or not they were in the correct posture. Minor verbal feedback was then provided to the subject, and the subject was allowed to assume his/her interpretation of the posture. However, in some cases the subject was assuming a posture that was grossly different from the requested posture. For example, in some cases the subject was instructed to assume a posture near 90° of knee flexion but actually positioned themselves in a posture near full flexion. In these cases, the researchers instructed the subject to adjust their posture.

Once the subject was in the posture of interest, the motion capture, force plate, and EMG data collection systems were initiated. The pressure system was then started and a researcher informed the subject and other researchers that data collection had begun. The researcher operating the pressure system collected data for 10 seconds, then stopped data collection and announced cessation to the subject and other researchers. Each researcher then reviewed their data for acceptability; the subject was instructed to be seated and place their knee near 90° making sure not to apply a load to the sensor. The sensor was then allowed to recover until the maximum and average pressure across the sensor was one pound per square inch or less. If an error occurred during data collection, the sensor was allowed to recover and the trial was repeated.

Data Analysis

EMG Data

Electromyography data was low-pass filtered to 500 Hz and high-pass filtered to 20 Hz using a 4th order Butterworth filter (MATLAB®; The MathWorks, Inc., Natick, MA). Recall, hardware filters were in place to high-pass filter to 10 Hz and low-pass filter to 500 Hz. The signal was then rectified by taking the absolute value. Data from all tests were normalized by dividing by the maximum voluntary contraction for each muscle. Mean amplitude values (MAVs) of the normalized values were then calculated by determining the running mean of every 102 samples which was 10% of the sampling rate [NIOSH 1992]. Trials containing evidence of artifacts were eliminated from the analysis. The lifting portion of the dynamic trials lasted approximately 2 seconds; however, time was normalized as a percentage of the lift for EMG analysis.

Data were analyzed using the Statistix 8.0 for Windows (Analytical Software, Tallahassee, FL). Residual analysis of the nontransformed EMG data indicated a fan-shaped pattern of residuals, thus the EMG data were transformed by taking the natural log, which resulted in a near normal distribution of residuals for all muscles. All analysis of variance (ANOVA) results reported below use the log-transformed data. Alpha levels were set at 0.05.

Pressure Data

Because the pressures applied to the knee sensor during the dynamic trials (i.e., trials where the subject twisted at the waist) exceeded the measurable range rendering these measurements inaccurate, only the static trials were evaluated for the pressure data. Additionally, squatting postures were not examined because the knee was not in contact with the ground during squatting. Therefore, pressure data from the static kneeling-near-90°-knee-flexion, kneeling-on-one-knee, and kneeling-near-full-flexion postures were evaluated. A cursory view of the data clearly indicated that the majority of pressure was transmitted to the knee via the patella and PTT, and. the total pressure being transmitted via these two structures was determined. It was then necessary to determine the ratio of the pressure that was being transmitted by the patella to that which was being transmitted by the PTT (please see Appendix B for detailed equations). For the static trials, all 10 seconds of data were evaluated.

Next, the magnitude of the pressures across the patella and PTT was determined as well as the mean pressure of both structures for every sample. This value was then summed and divided by the total number of samples. This yielded the mean value of the average pressure on the patella and PTT across all samples, or the mean of the mean pressure. The mean maximum pressure was also determined by summing the maximum pressures at each sample and dividing by the total number of samples.

Finally, some measurement of the distribution of pressure across the structures was necessary; therefore, a measure of variance was taken. The variance was calculated for the patella and PTT for each sample. These variances were then summed and divided by the total number of samples yielding the mean variance across each structure.

Statistical analyses were performed using Statistix 8.0 for Windows. Analyses performed included a split-plot ANOVA and a priori orthogonal contrasts. Contrasts for kneepad conditions included: 1) comparing the no-kneepad and kneepad-present conditions, and 2) comparing nonarticulated and articulated kneepads. Contrasts for posture states included: 1) comparison of kneeling with both knees in full flexion to kneeling on the right knee only (across both heights), and 2) comparing the 38 in work height to the 48 in work height. All contrasts were tested using a *t* statistic with an alpha level of 0.05. As an exploratory analysis, multiplicity corrections were not applied in the data analyses [Bender and Lange 2001]. Specifically, a comparison-wise Type I error rate alpha level of 0.05 was employed.

Net Forces and Moments and Resulting Joint Kinematics

To determine the right knee angles and net externally applied forces and moments at the right knee, a computational model was developed in the MATLAB® software (The Mathworks Inc., Natick, MA) on a personal computer. This model is based on inverse dynamics [Winter 1990] which uses measured ground reaction forces and anthropometric measurements to estimate reaction forces and moments. In this linear model, the upper leg (femur) and the lower leg (tibia, fibula, and foot) are modeled as rigid bodies attached via a pin joint with three rotational degrees of freedom. Using the anatomical marker set (described previously), an anatomical shank coordinate system (ASCS) was constructed. This coordinate system was oriented such that when the subject was standing in standard anatomical position, the x-axis was in the medial-to-lateral direction, the y-axis was in the posterior-to-anterior direction, and the z-axis was in the inferior-to-superior direction. The origin of this system was at the right knee joint center. Similarly, an anatomical coordinate system was created for the thigh, with the origin at the right hip joint center. These coordinate systems were established to move with the subjects as they assumed the various postures, thereby relating all postures to clinical measurements. A positive moment about the x axis represents extension rotation, the y axis represents varus rotation, and the z axis represents internal rotation.

The developed computational model is based on the following assumptions.

- *The knee is assumed to be a frictionless pin joint. This allows all forces to pass directly through the joint center.*

- *Segments are assumed to be rigid with mass concentrated at the center of mass locations. This allows one center of mass to represent the weight of the segment.*

- *The linear relationship between external forces and moments is applied to the knee. This allows a 3-D model to be used to determine the external forces and moments applied to the knee.*

- *The relative movement of pelvic bones is negligible. This allows approximation of the hip joint center from palpable pelvic landmarks.*

- *Thickness of subcutaneous tissue between bone and skin is minimal. This allows the assumption that markers placed on palpable landmarks are directly located on the landmark.*

- *The measured thigh-calf and heel-gluteus contact forces are concentrated at the measured center of pressure location. This allows the contact forces to be represented as a single resultant force, opposed to a pressure distribution.*

- *The effect of patellar tendon and tibial tubercle on forces externally applied to the tibia is negligible. This assumption allows ground contact forces measured at the ground-knee or ground-kneepad interface to be assumed to act at some distance away from the knee joint center and not be affected by the patellar tendon or tibial tubercle.*

NOTE: Please see Appendix A for detailed information regarding the development of coordinate systems used in the above described model.

Joint angles were determined for each posture using Euler angle decomposition. The largest joint rotations occurred about the x-axis (extension/flexion), followed by the y-axis (varus/valgus), and the z-axis (internal/external rotation), yielding an Euler order of Xy'z''. The rotation matrix from the anatomical thigh to the anatomical shank coordinates was created to determine the Euler angles. Therefore, motion of the thigh was with respect to the shank. The included angle was defined as the angle between the thigh and shank along the flexion/extension axis. Varus rotation was positive along the y axis, and internal rotation was positive along the z axis.

Ground reaction forces, segment weight, thigh-calf contact force ($F_{t/c}$), and heel-gluteus contact force ($F_{h/g}$) were inputs into the computational model. External force diagrams for kneeling near full flexion with respect to the global coordinate system (GCS), coordinate system of the laboratory constructed by the motion capture system, and the ASCS are shown in Figure 6 and 7. The center of mass location and weight of the shank+foot were determined using equations from Hinrichs [1990] which were adjusted to use the knee joint center and ankle joint center in this model. The reaction forces and moments were assumed to act in the positive directions.

Squatting and kneeling create a contact force between the thigh and the calf ($F_{t/c}$). Kneeling near full flexion also creates this contact, and in some subjects there is additional contact between the heel and the gluteal muscles ($F_{h/g}$). Both $F_{t/c}$ and $F_{h/g}$ were modeled as resultant forces whose lines of action were in the anterior direction of the shank with centers of pressure some distance (i.e., measured by the pressure sensor) along the long axis of the shank, thereby contributing only to extension moments and anterior forces. Forces at the foot (F_1), forces at the knee (F_2), and the weight of the lower leg (W_{LL}) were measured with respect to the GCS and transformed into the ASCS (Figure 6). Thigh-calf and heel-gluteus contact forces were measured with respect to the ASCS.

Figure 7 shows the orientation of the forces and moments as presented in this research, with respect to the ASCS. External force diagrams for all postures are shown in Figure 8.

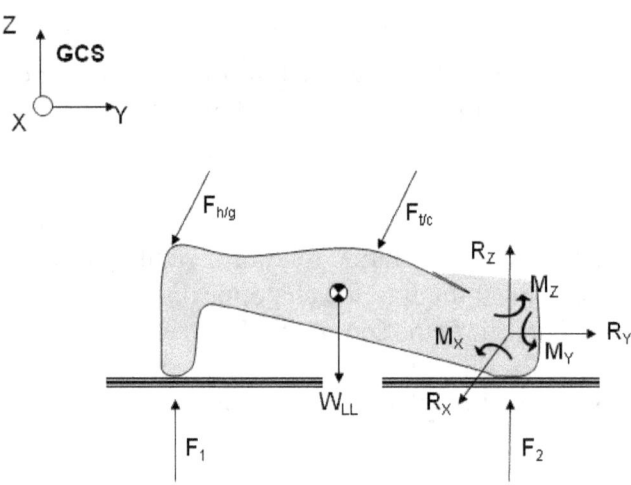

Figure 6. Diagram of external shank forces and reaction forces and moments for kneeling near full flexion with respect to the global coordinate system (GCS).

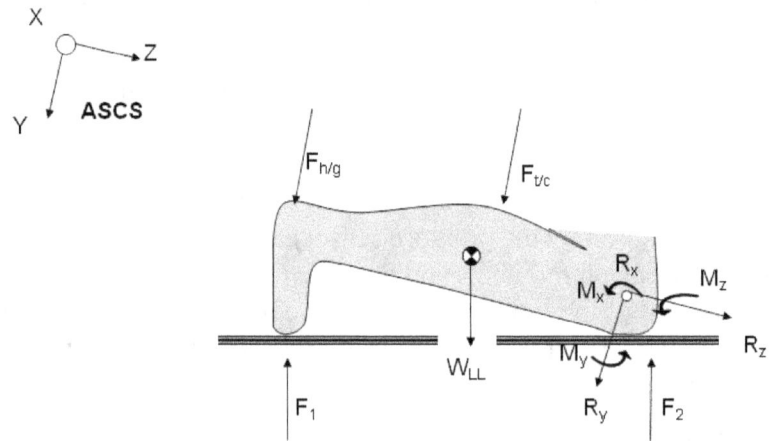

Figure 7. Diagram of external shank forces and reaction forces and moments for kneeling near full flexion with respect to the anatomical shank coordinate system (ASCS).

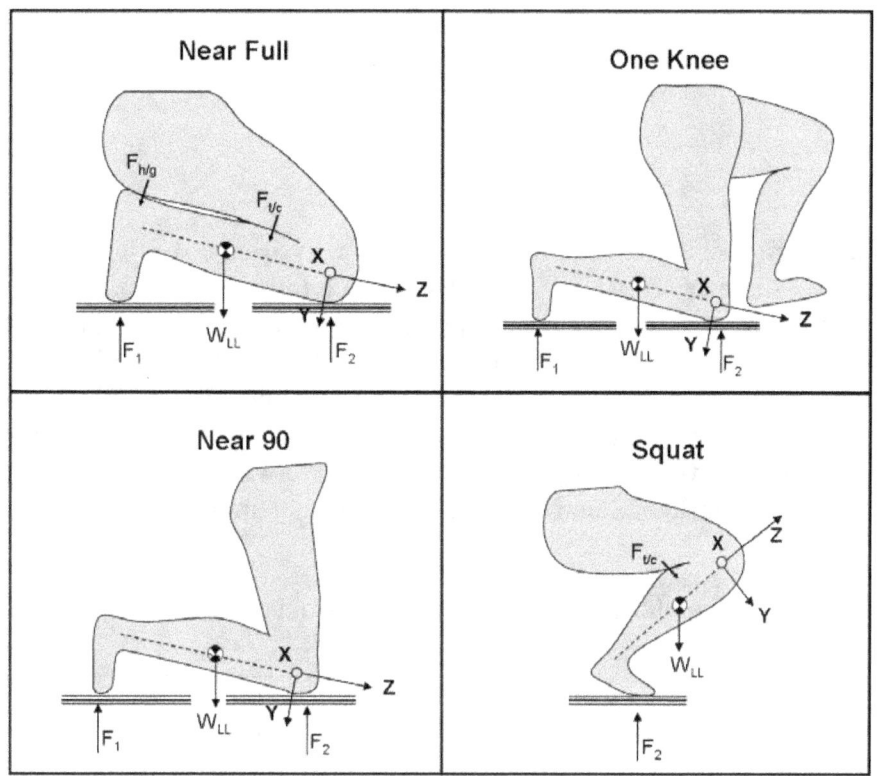

Figure 8. External force diagrams with respect to the anatomical shank coordinate system (ASCS).

Joint equilibrium was assumed; therefore external forces in the x, y, and z directions were all summed to zero to determine the reaction force, R, necessary to stabilize the knee due to the application of these external forces. The sum of the external moments at the knee joint in the x, y, and, z directions were also summed to equal zero, and subsequently the reaction knee moment was then determined. The data presented in this study represents the net externally applied forces and moments. The force necessary to stabilize the knee from the application of these forces and moments will be equal in magnitude and opposite in direction to the forces and moments presented in this study.

Statistical analyses were performed using Statistix 8.0 for Windows. A two-way (3 kneepads × 4 postures) split-plot ANOVA was performed to determine if significant differences existed in knee angles, forces, and moments between postures and kneepad conditions. A priori orthogonal contrasts included comparisons of the no-kneepad to kneepad-present (articulated and nonarticulated) conditions, high-flexion (squatting and kneeling near full flexion) and low-flexion (kneeling near 90° knee flexion) postures, and squatting versus kneeling (kneeling near 90° knee flexion, kneeling on one knee, and kneeling near full flexion) postures. All orthogonal contrasts were tested using a t statistic with an alpha value of 0.05. A priori nonorthogonal contrasts were tested to determine if significant differences existed between the kneepad conditions using Scheffe F-test ($p < 0.05$).

Results

EMG Data

Recall that static trials were excluded from this analysis since only minimal muscle activity was observed. A summary of ANOVA results for each muscle is provided in Table 1. As can be seen in this table, 9 out of 10 muscles investigated were affected by an interaction between the posture adopted and the location of the block throughout the lift. The sole exception was the left vastus lateralis, which was affected by the main effects of posture and block position, but not their interaction. The kneepad main effect was not found to influence EMG activity for any of the muscles studied; however, a significant interaction between kneepad condition and block position was found to influence activity of the left biceps femoris.

Table 1. Summary of significant main effects and interactions for normalized activity of all muscles.

	Kneepad (A)	Posture (B)	A*B	Block position (C)	A*C	B*C	A*B*C
Left vastus lateralis		***		***			
Left rectus femoris		***		***		***	
Left vastus medialis		***		***		***	
Left biceps femoris				***	*	**	
Left semitendinosus		**		***		***	
Right vastus lateralis		***		***		***	
Right rectus femoris		***		***		***	
Right vastus medialis		***		***		***	
Right biceps femoris		***		***		***	
Right semitendinosus		***		***		***	

* indicates significant difference with *$p<0.05$, **$p<0.01$, ***$p<0.001$

The location of the block was measured throughout the lifting trials. The angular position of the block with respect to its starting position was determined. The subject lifted the block which was located directly to his/her right (0°) and transferred it to a location directly to his/her left (180°). Figure 9 provides summaries of EMG data at these two positions (as the block was picked up and then set down) and at three intermediate points during the lift (45°, 90°, and 135°), providing an overall representation of the interaction of posture and block location.

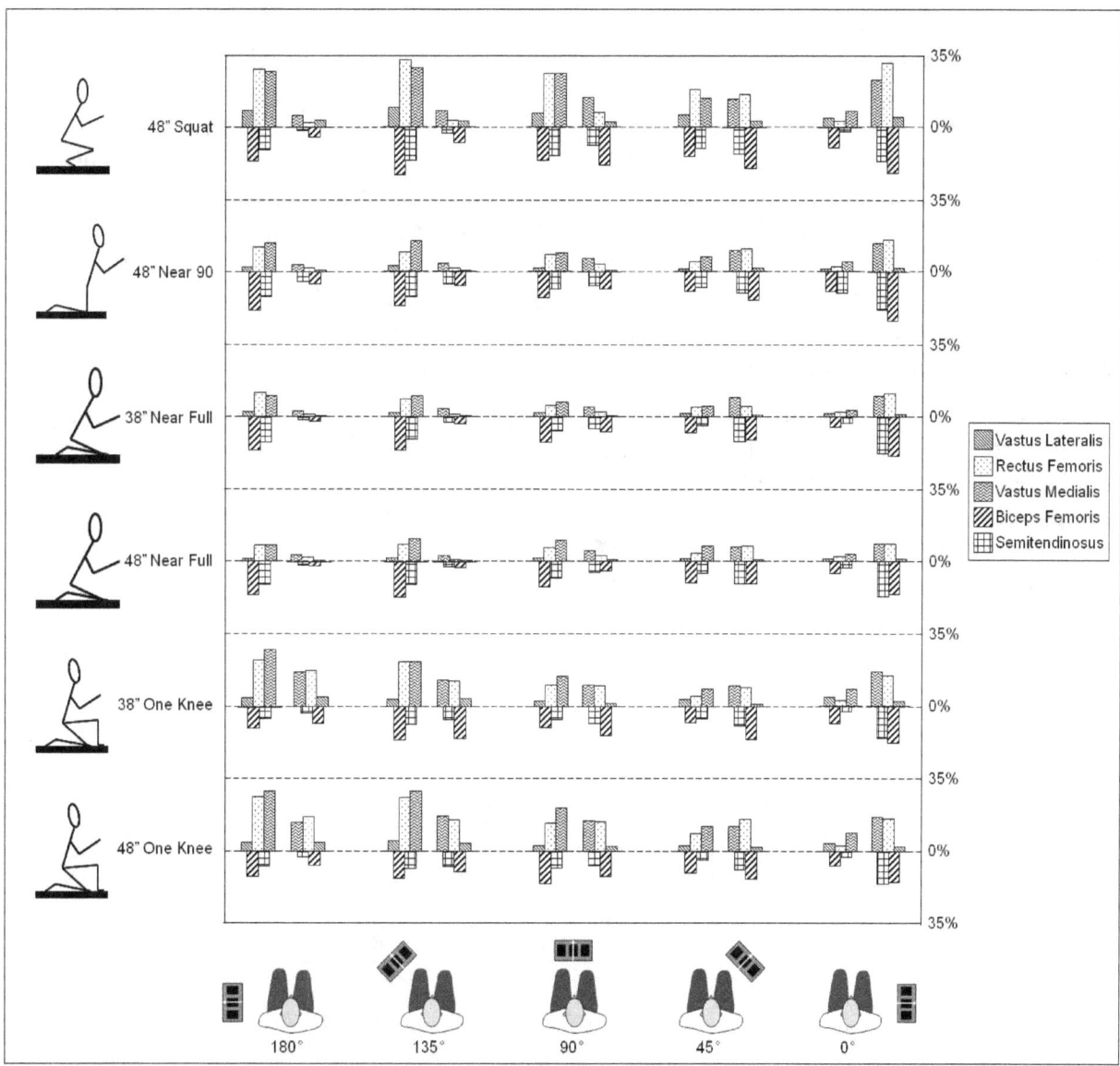

Figure 9. Summary of EMG activity by posture and block position. Each posture and block position has two groups of bars representing the activity of the left and right thigh muscles. Note that the bars representing the left and right thigh muscles mirror their arrangement in the body if looked at from the superior aspect.

Inspection of Figure 9 provides several insights into the activation of the thigh muscles in different postures and at different phases during the lift. One consistent pattern that can be observed with all postures is the high activation of right thigh muscles at the beginning of the lift and the high activation of the left thigh muscles as the load is transferred to the left side. Peak activity of the left thigh muscles tended to occur when the block was at the 135° (front left) or 180° (left) positions. In most cases, the lowest overall EMG activity was observed when the block was at the 45° (front right) or 90° (front) position.

Several differences in muscle activity can be noted between postures in Figure 9. Squatting generally exhibited the highest muscle activity levels throughout the entire lift, and this was particularly true in the knee extensors. At the start of the lift (0° condition), the right knee extensors exhibited more than twice the activity when squatting compared to any other posture (30% MVC compared to 15% MVC in kneeling near 90° knee flexion, the next highest posture). Kneeling near full flexion (in either 38-in or 48-in seam heights) typically exhibited the lowest muscle activity of any posture. Kneeling near 90° knee flexion exhibited very similar activation patterns to the kneeling-near-full-flexion condition, but tended to have slightly higher EMG activity. Kneeling on one knee (either seam height), however, resulted in higher activity than the other kneeling postures, with a notable increase in right and left side extensor EMG at the end of the lift compared to the other kneeling postures.

The presence or absence of kneepads had virtually no effect on EMG activity in this study. The one exception is that the left biceps femoris was affected by an interaction of kneepad and block position, shown in Figure 10. Compared to the two kneepad-present conditions, the no-kneepad condition resulted in lower EMG activity of the left biceps femoris in all block positions with the exception of the initial position (0°).

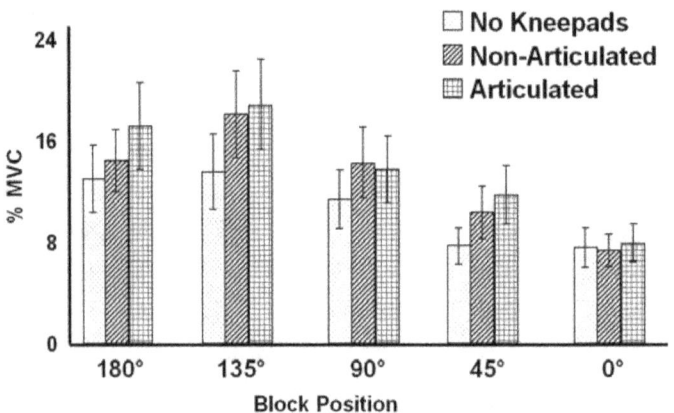

Figure 10. Interaction of kneepad condition and block position on EMG activity of the left biceps femoris across postures.

Pressure Data

For all postures tested, the majority (>60%) of the pressure was placed on the PTT region (Figure 11). No significant difference existed between the no-kneepad condition and the two kneepad-present conditions; however, a significant difference existed within the two kneepad-present conditions with the articulated kneepad exhibiting a greater mean pressure ratio for the PTT region ($p < 0.0001$; 86% versus 74%). When evaluating the effect of the postures, the kneeling-near-full-flexion postures showed a significantly greater mean pressure ratio for the PTT region ($p < 0.0001$; 88% versus 74%) when compared to the kneeling-on-one-knee postures. No significant difference between the 38 in and 48 in working heights was observed for any dependent measures.

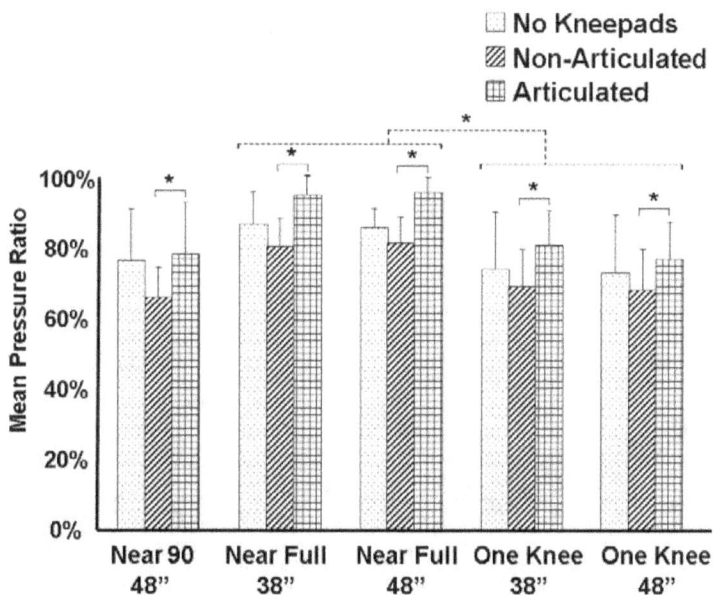

Figure 11. Mean pressure ratio at the PTT region across postures. (* indicates significant difference with $p < 0.05$).

The mean of the mean pressure at the patella region showed that only modest amounts of pressure (<10 psi) were applied to the patella for all postures (Figure 12). As with the pressure ratio, it was found that no significant difference existed between the no-kneepad condition and the two kneepad-present conditions, but a significant difference existed between the two kneepad-present conditions with the nonarticulated kneepad exhibiting a greater mean of the mean pressure for the patella region ($p < 0.0001$; 4.3 versus 2.4 psi). When evaluating the effect of the postures, it was determined that the kneeling-on-one-knee postures showed significantly greater mean of the mean pressure for the patella region ($p < 0.0001$; 5.1 versus 0.9 psi) when compared to the kneeling-near-full-flexion postures.

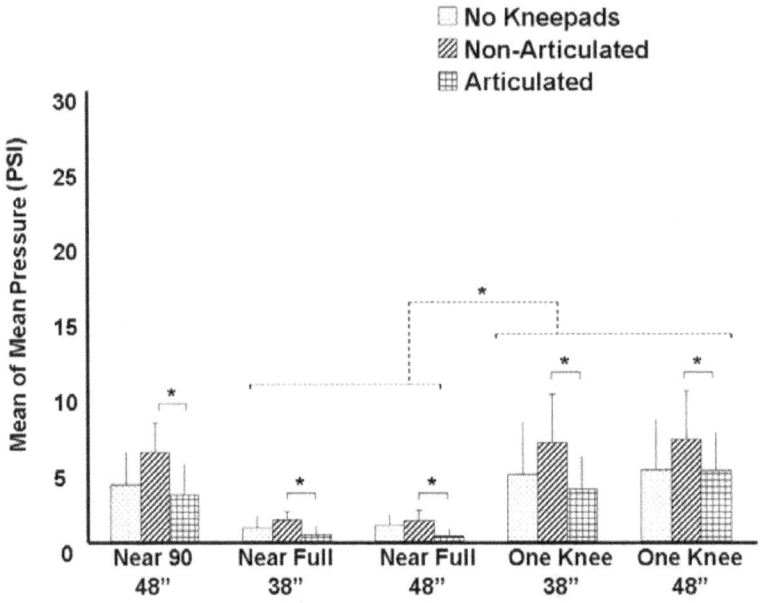

Figure 12. Mean of the mean pressure at the patella region for the various postures. (*indicates significant difference with $p < 0.05$).

The mean of the mean pressure at the PTT region showed a considerably higher level of applied pressure (>15 psi) than that which was observed for the patella region (Figure 13). As with the mean of the mean pressure for the patella region, a significant difference ($p < 0.05$) was observed for posture and subject. In contrast to the patella region, the PTT region did not show significant differences due to the kneepad-present condition. As with the patella region, the kneeling-on-one-knee postures showed significantly greater mean of the mean pressure ($p < 0.0001$; 17.9 versus 8.2 psi) when compared to the kneeling-near-full-flexion postures.

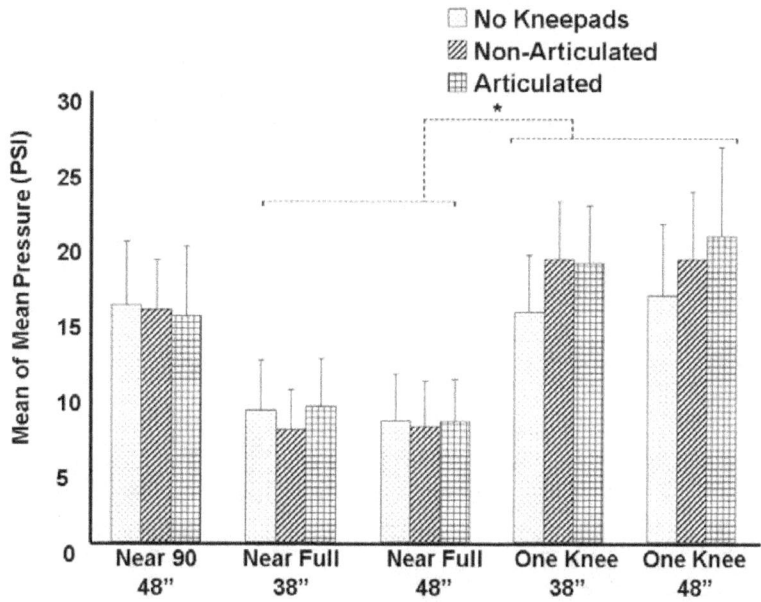

Figure 13. Mean of the mean pressure at the PTT region for the various postures. (*indicates significant difference with $p < 0.05$).

The mean of the maximum pressure at the patella region showed that highly variable (1.3±1.1 to 27.1±17.2 psi) maximum pressures were applied to the patella for the different postures (Figure 14). Again, no significant difference existed between the no-kneepad condition and the two kneepad-present conditions, but a significant difference did exist between the two kneepad-present conditions with the nonarticulated kneepad exhibiting a greater mean of the maximum pressure for the patella region (p = 0.0006; 15.2 versus 9 psi). Additionally, the kneeling-on-one-knee postures showed significantly greater mean of the maximum pressures for the patella region (p < 0.0001; 18.4 versus 3.1 psi) when compared to the kneeling-near-full-flexion postures.

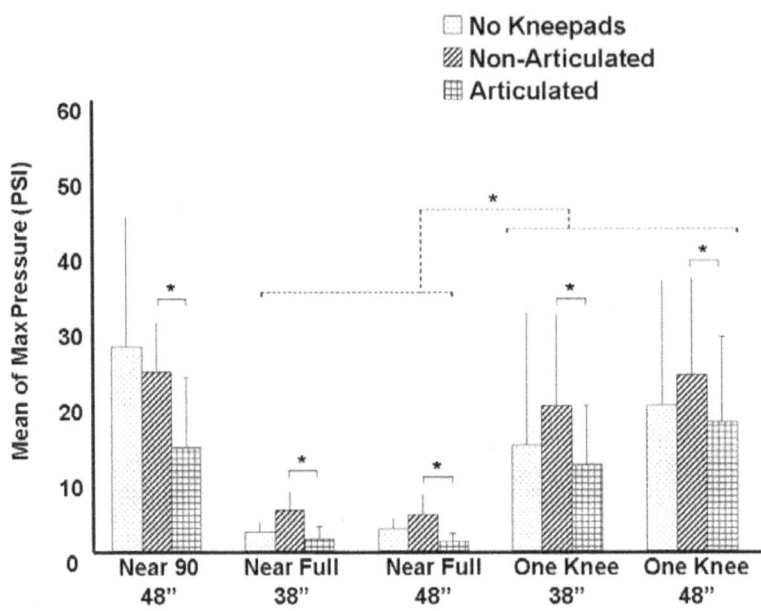

Figure 14. Mean of the maximum pressure at the patella region for the various postures. (*indicates significant difference with p < 0.05).

The mean of the maximum pressure at the PTT region showed a considerably higher level of pressure (>25 psi) than that observed for the patella region for all postures regardless of the kneepad condition (Figure 15). No significant difference existed between the two kneepad-present conditions, but a significant difference existed between the no-kneepad and the two kneepad-present conditions with the no-kneepad condition exhibiting a greater mean of the maximum pressure for the PTT region ($p < 0.0001$; 41.9 versus 28.5 psi). Looking at significant differences within posture, the only significant difference was found between the kneeling-near-full-flexion and kneeling-on-one-knee postures, with the kneeling-on-one-knee postures exhibiting a greater mean of the maximum pressure ($p < 0.0001$; 37.1 versus 28.7 psi).

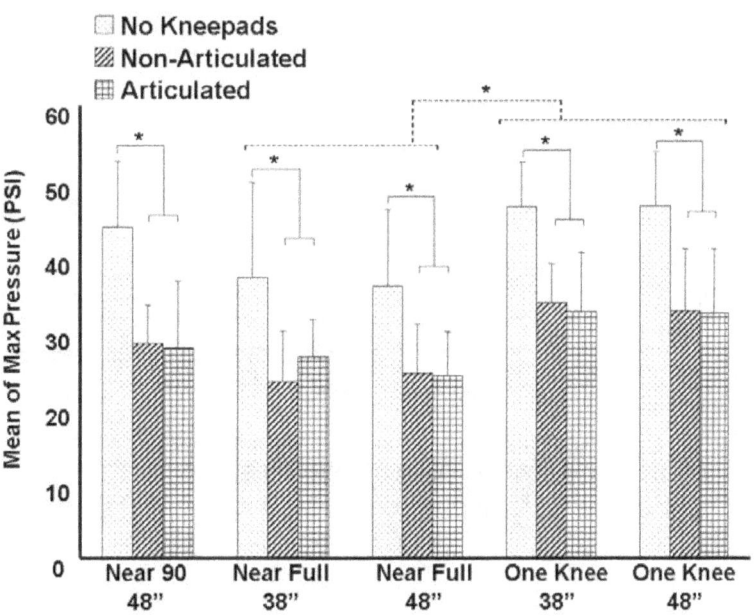

Figure 15. Mean of the maximum pressure at the PTT region for the various postures. (*indicates significant difference with $p < 0.05$).

The mean of the variance at the patella region varied considerably across postures (Figure 16). The kneeling-on-one-knee postures showed significantly greater means of the pressure variances for the patella region ($p < 0.0001$; 39.3 versus 1.1 psi) when compared to the kneeling-near-full-flexion postures.

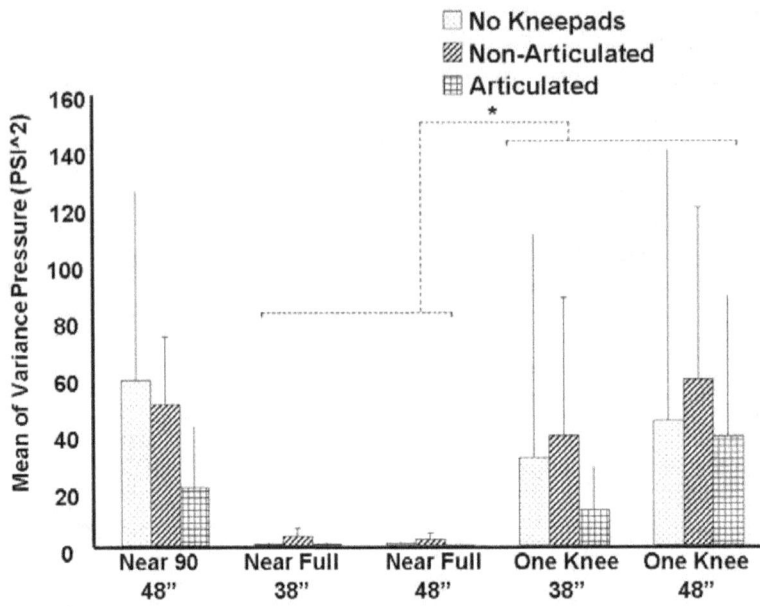

Figure 16. Mean of the variance at the patella region for the various postures (*indicates significant difference with $p < 0.05$).

The mean of the variance at the PTT region showed relatively consistent levels of variance across postures (Figure 17). The no-kneepad condition was found to have a significantly greater mean of the variance for the PTT region when compared to the two kneepad-present conditions ($p < 0.0001$; 173.3 versus 67.6 psi). No significant difference was found between the two kneepad-present conditions. The kneeling-on-one-knee postures showed significantly greater means of the variances for the PTT region ($p < 0.0001$; 124.2 versus 85.8 psi) when compared to the kneeling-near-full-flexion postures.

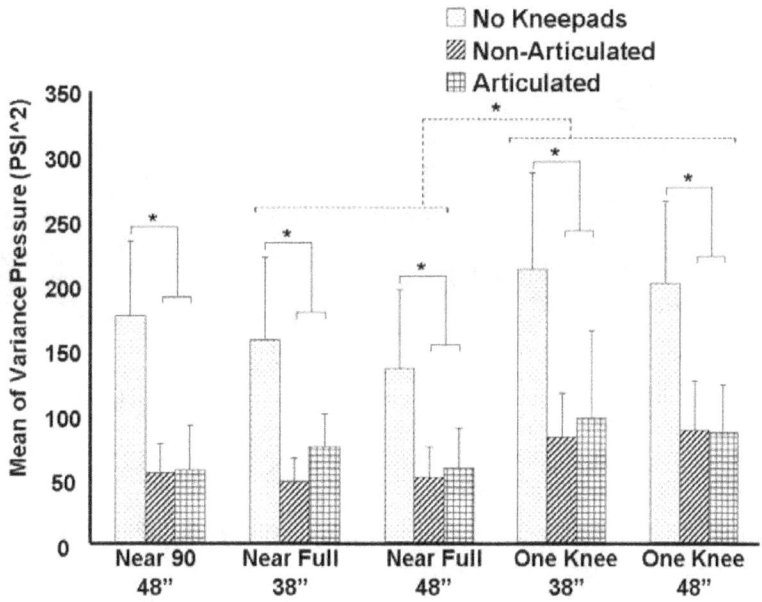

Figure 17. Mean of the variance at the PTT region for the various postures (*indicates significant difference with $p < 0.05$).

Net Forces and Moments and Resulting Joint Kinematics Data

All subjects had thigh-calf contact forces greater than 20% of body weight (BW) for squatting with a mean of 39 ±14 % BW. Mean thigh-calf contact when kneeling near-full flexion was 28 ± 13% BW. In the kneeling-near-full-flexion posture, seven subjects had contact between the heel and gluteal muscles, with a mean of 11 ± 6% BW. Typical pressure distribution profiles for squatting and kneeling near full flexion are shown in Figure 18.

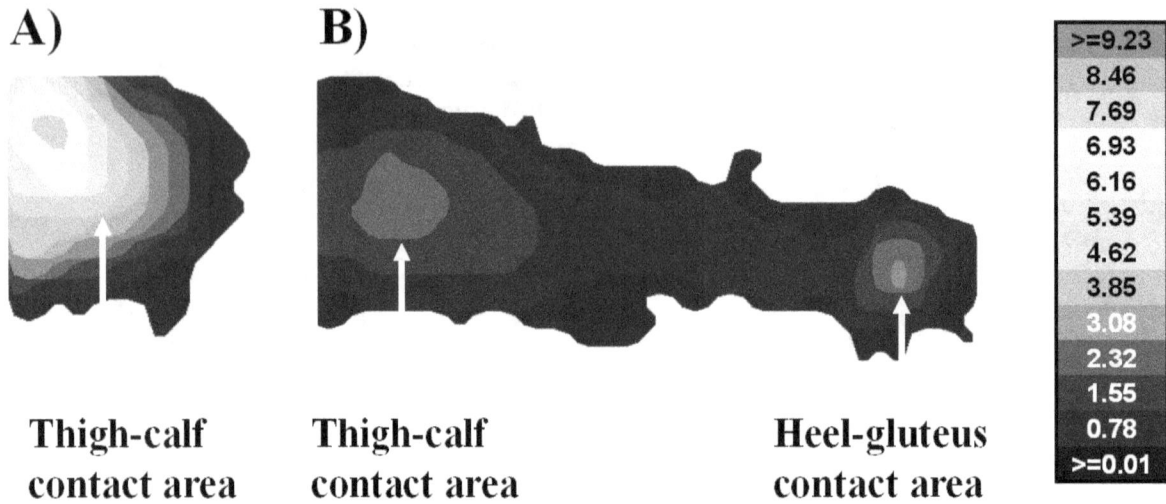

Figure 18. Typical pressure distribution (psi) for squatting (A) and kneeling near full flexion (B).

Significant differences were seen between postures for the included ($p < 0.001$) and internal rotation angles ($p < 0.001$) (Figure 19). When compared to the kneeling postures, squatting showed higher flexion and internal rotation ($p < 0.001$). A priori orthogonal contrasts showed significant differences in knee flexion between the high-flexion (squatting, kneeling near full flexion) and low-flexion (kneeling on one knee, kneeling near 90° of knee flexion) postures ($p < 0.001$), and between kneeling near 90° of knee flexion and kneeling on one knee ($p < 0.001$). Differences in internal rotation angles were observed between high-flexion and low-flexion postures ($p < 0.001$), and between squatting and kneeling near full flexion ($p < 0.001$) postures. The no-kneepad condition showed higher knee flexion (included angles = 35 ± 24°) than the nonarticulated kneepad-present condition (included angles = 41 ± 26°) ($p = 0.033$). No significant differences were observed between the articulated (40 ± 26°) and nonarticulated kneepads.

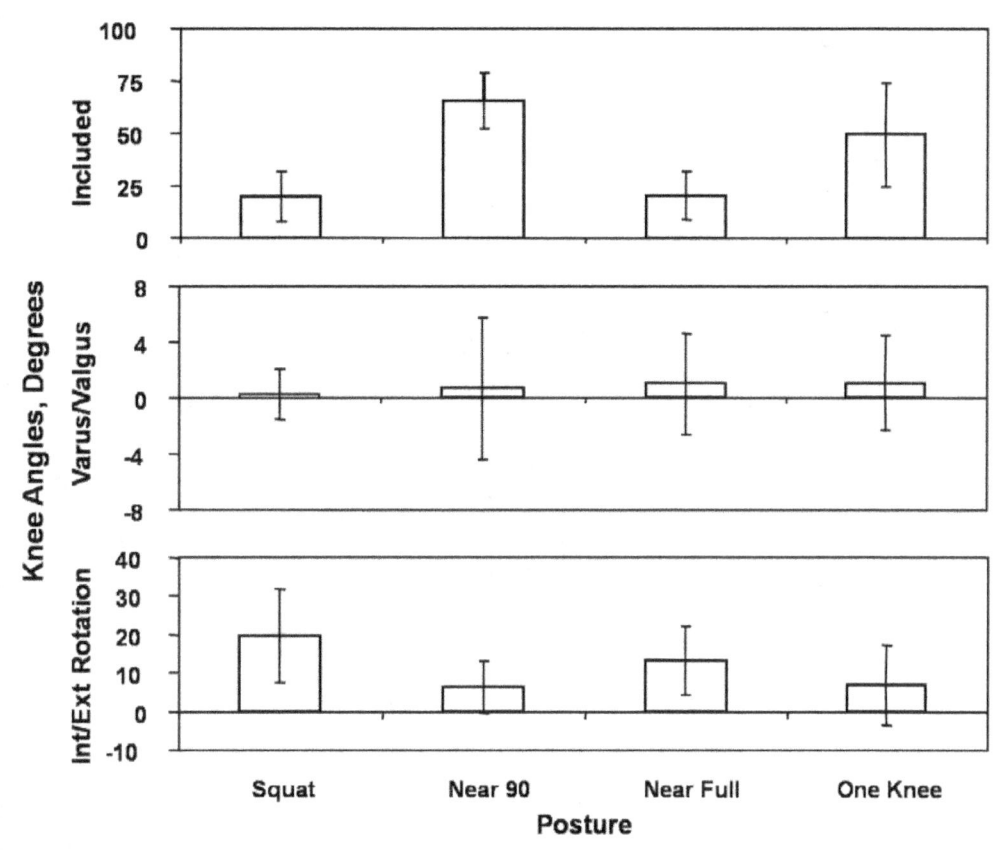

Figure 19. Knee angles (degrees) for all postures. Varus and internal rotation angles are positive.

Posture had a significant effect on tibial forces (Figure 20). When compared to the kneeling postures, squatting showed significantly higher superior, lower posterior, and lower resultant forces ($p < 0.001$). Low-flexion postures showed greater posterior, inferior, and resultant forces than the high-flexion postures ($p < 0.001$). Squatting showed significantly higher medial ($p = 0.002$), superior ($p < 0.001$), and resultant forces ($p < 0.001$), but lower posterior forces than kneeling near full flexion ($p < 0.001$). Kneeling on one knee showed significantly higher medial, posterior, inferior, and resultant force magnitudes than kneeling near 90° knee flexion ($p < 0.001$). The use of kneepads affected the medial forces ($p = 0.036$), and the no-kneepad condition (-6.1 ± 8.1 % BW) had higher medial forces than the articulated kneepad-present condition (-4.3 ± 7.6 % BW).

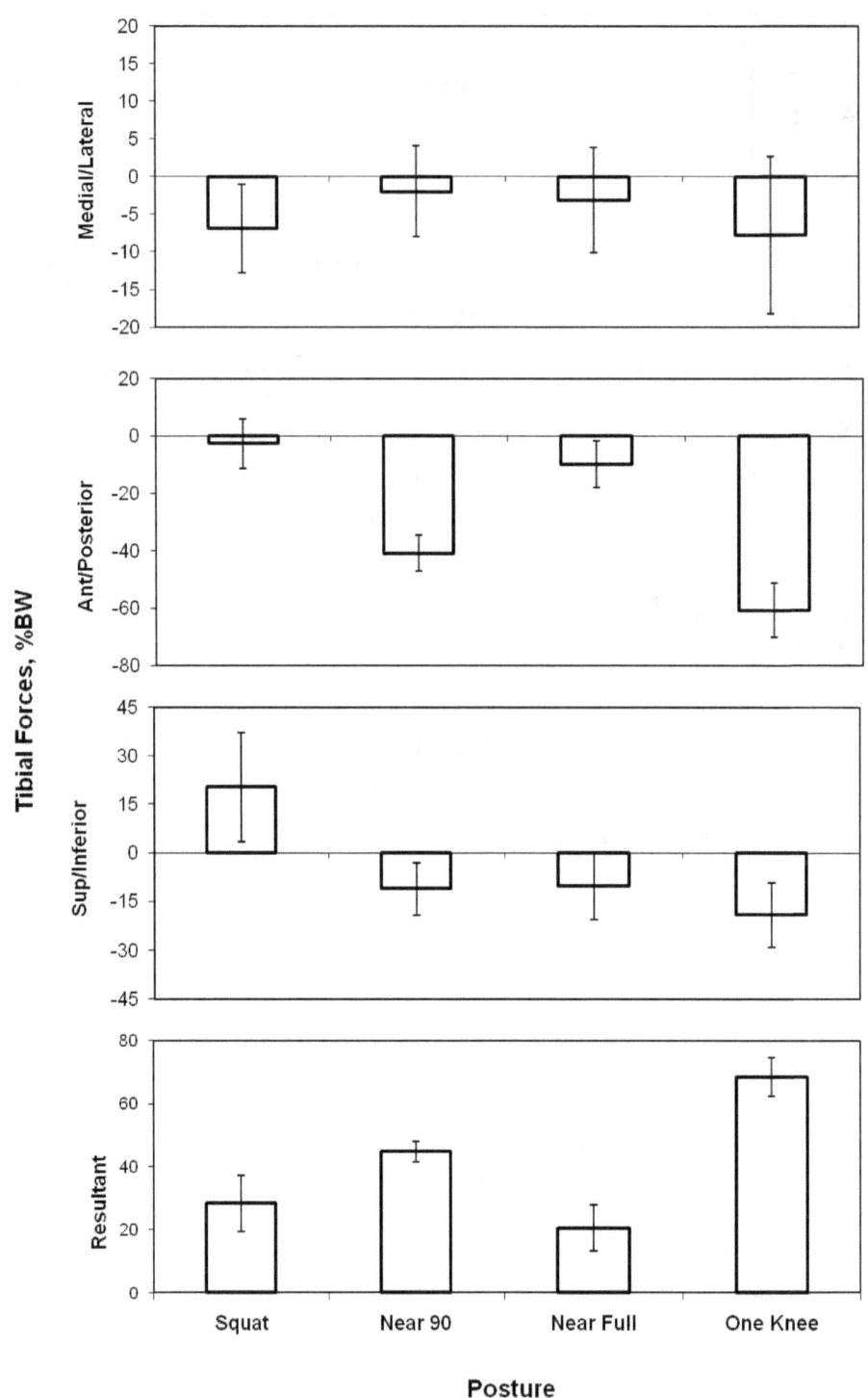

Figure 20. Mean tibial forces normalized to body weight (% BW). Lateral, superior, and anterior forces are positive.

Posture also had a significant affect on tibial moments (Figure 21). Flexion moments had the greatest magnitudes for all postures. Of the postures examined, kneeling on one knee had the greatest flexion moments, squatting had the greatest varus moments, and kneeling near 90° knee flexion had the greatest internal rotation moments. Kneeling on one knee and squatting had the highest resultant moments with comparable magnitudes of 5 ± 4% BW*Ht for kneeling on one knee and 5 ± 2% BW*Ht for squatting. When compared to the kneeling postures, squatting showed significantly lower internal rotation ($p = 0.027$), increased varus ($p < 0.001$), and increased resultant moments ($p = 0.027$). High-flexion postures showed significantly higher varus ($p < 0.001$), internal rotation ($p = 0.009$), and resultant moments compared to the low-flexion postures ($p = 0.007$). Squatting showed significantly higher varus moments than kneeling near full flexion ($p < 0.001$). Kneeling on one knee showed significantly higher flexion ($p < 0.001$), varus ($p = 0.012$), and resultant moments, but significantly lower ($p < 0.001$) internal rotation moments than kneeling near 90° knee flexion ($p = 0.01$). Kneepad use affected the internal rotation moments ($p < 0.001$), with the no-kneepad condition (0.23 ± 0.32 % BW*Ht) having significantly smaller internal rotation moments than the nonarticulated (0.36 ± 0.36 % BW*Ht) ($p = 0.049$) and articulated kneepad-present conditions (0.36 ± 0.35 % BW*Ht) ($p = 0.008$).

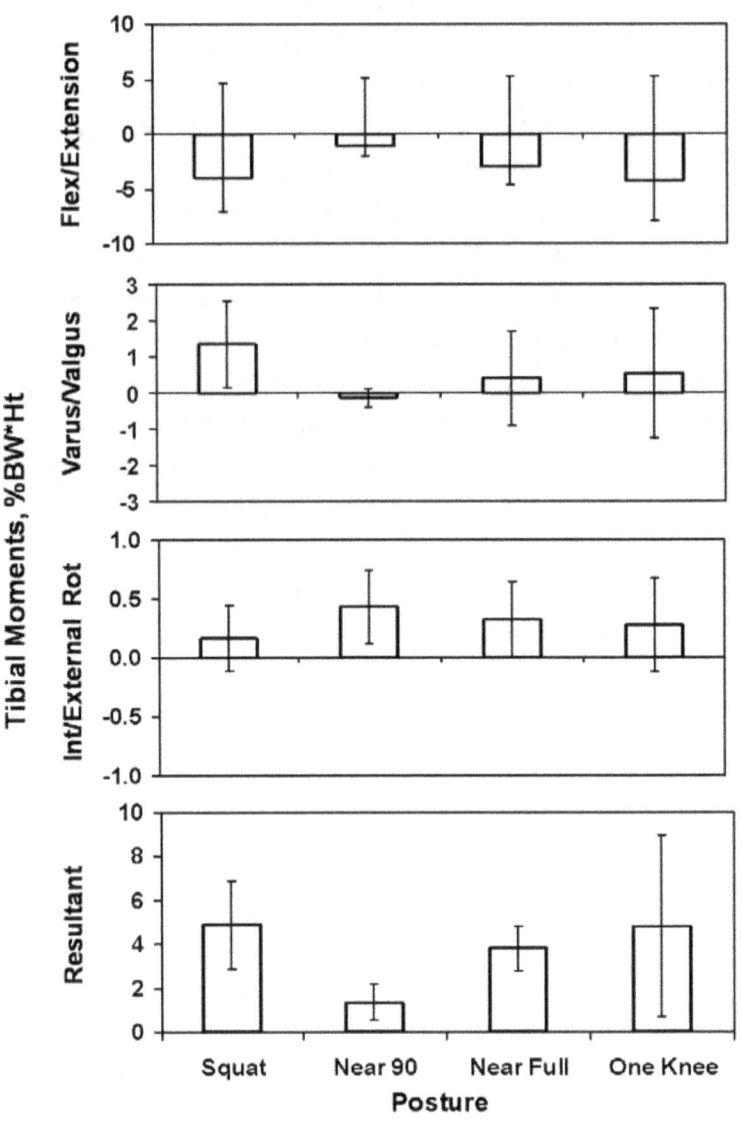

Figure 21. Mean tibial moments normalized to body weight times height (% BW*Ht). Extension, varus, and internal rotation moments are positive.

Discussion of Results

The results of this study include several key findings which may be useful in providing recommendations to reduce the prevalence and severity of knee injuries in low-seam coal mine workers. The three biomechanical parameters investigated demonstrated that, relative to oneanother, the postures in this study had potential beneficial and detrimental effects on the worker. The key points of interest for each parameter are as follows:

EMG

The flexors and extensors of the knee were affected by the interaction between the posture adopted and the location of the block during the lift. The lowest EMG activity was typically observed when the block was raised from the ground and in front of the subject. This suggests that supplies should be stored on elevated platforms in the mine such that minimal leg muscle activity is required to load and unload supplies. The highest muscle activity was noted for squatting followed by kneeling on one knee. Thus, both of these postures should be expected to lead to an earlier onset of fatigue for the flexors and extensors of the knee than kneeling near full flexion and kneeling near 90° of knee flexion.

Pressure

The majority of the pressure was found to be transmitted to the knee via the PTT (over 60%). Kneepads were not found to have a significant effect on the pressure ratio between the patella and the PTT, the mean of the mean pressure, or the mean of the maximum pressure. Therefore, contrary to the original hypothesis of this study, kneepads, in their current form, may result in increased comfort but have no measurable ability to reduce the risk of pressure-driven knee injuries. However, the hypothesis that postures would have a significant effect on pressures at the knee was supported by this study. Compared to kneeling on one knee, kneeling near full flexion resulted in a significant shift in the distribution of the pressure between the patella and the PTT regions with more pressure being applied to the PTT region. This shift was also found to correlate with an increase in the mean pressure applied to the PTT region. The mean pressure at the PTT region was significantly greater than that for the patellar region. Due to the minimal muscle activity required to maintain a kneeling-near-full-flexion posture, the assumption is that mine workers would frequently and naturally adopt this posture to reduce the onset of fatigue. However, pressure data clearly indicate that kneeling near full flexion places significant pressure on the PTT region which may lead to injuries such as prepatellar bursitis. It is imperative that manufacturers redesign kneepads such that they effectively reduce the pressure at the knee and, specifically, at the PTT region.

Net Forces and Moments

As hypothesized, significant differences were found between squatting and kneeling postures as well as between the high-flexion (squatting and kneeling near full flexion) and low-flexion (kneeling near 90° knee flexion and kneeling on one knee) postures. Both high-flexion kneeling and squatting create a significant change in internal joint structure orientations, such as femoral rollback and varying segment contact forces. During increased knee flexion, the tibia internally rotates. However, tibial loading during squatting is significantly different during kneeling, with mean forces acting in the medial, posterior, and superior directions of the tibia when squatting. In all the kneeling postures examined, the forces acting on the tibia occurred in the medial, posterior, and inferior directions. When kneeling on one knee, the highest resultant force magnitudes were noted and shear loading across the knee joint (i.e., medial forces) occurred. In effect, high posterior forces, as much as 60% BW are transmitted to the tibia in addition to flexion moments up to 5% BW*Ht. These data suggest that statically kneeling on one knee results in osteoarthritis and mensical tears. Furthermore, the high flexion and loading patterns present when kneeling near full flexion and squatting may also place a worker at a higher risk for developing osteoarthritis of the medial compartment and meniscal tears.

Implications of Results

Preliminary investigations demonstrated that the muscle activity of the flexors and extensors of the knee was minimal when statically kneeling, which shows that the least amount of activity was observed for kneeling near full flexion. Additionally, the mean pressure at the knee was less when kneeling near full flexion than in other postures. This suggests that workers would likely experience less fatigue and more comfort when kneeling near full flexion. This posture is likely a predominant posture used by workers where the vertical working height is restricted. However, this posture resulted in increased internal tibial rotation, which could lead to osteoarthritis or meniscal tears. As such, prolonged use of this posture is not advised.

Lateral lifting tasks are very common in the low-seam mining industry (e.g., building stoppings, shoveling). For these tasks, the muscle activity of the flexors and extensors of the knee are expected to be at their greatest when squatting or kneeling on one knee. This suggests that these two postures may result in an earlier onset of fatigue. However, the resulting pressure at the knee was not evaluated in this study as the magnitude of these pressures exceeded the sensor capacity. NIOSH has recently published a study whereby these pressures were estimated for the postures investigated in the current study while performing a lateral lift [Mayton, et al. 2010]. These data indicated that kneeling near full flexion while performing a lateral lift resulted in pressures that were six times greater than statically kneeling near full flexion. Furthermore, the estimated pressure on the knee was greater for kneeling near full flexion than for kneeling near 90° of knee flexion and kneeling on one knee. Therefore, for lateral lifting tasks, kneeling near full flexion is not advised. Rather, workers should attempt to adopt a kneeling-near-90°-of-flexion posture when physically possible. Alternatively, kneeling on one knee may be adopted; however, this will likely result in an earlier onset of fatigue. Furthermore, statically kneeling on one knee resulted in high shear forces which are likely to still exist while performing a lateral lift.

Kneepads were not found to significantly reduce mean pressures applied to the knee for any posture where the knee was in contact with the ground. Although squatting does remove pressure from being applied to the knee, it is recommended that this posture should generally be avoided. Statically squatting resulted in high moments at the knee which, similar to kneeling near full flexion, could result in osteoarthritis or meniscal tears. Futhermore, squatting while performing a lateral lifting task resulted in large amounts of muscle activity which will likely lead to an earlier onset of fatigue. While statically kneeling on one knee, high shear loading was observed, making it likely that these shear forces are present when performing a lateral lift as well. Therefore, kneeling near 90° of knee flexion should be the primary choice for performing lateral lifting tasks, with kneeling on one knee as the secondary choice. When kneeling on one knee is necessary, mine workers should alternate the knee in contact with the ground. This posture, however, is not recommended for static tasks.

Based on the data from the current study, several recommendations are made:

1. Kneeling near full flexion is likely to be frequently adopted in environments where the vertical height is restricted; however, workers should avoid using this posture for prolonged periods of time. Rather, they should attempt to alternate between kneeling near full flexion and a posture that does not place the knee at a high risk for osteaoarthritis and meniscal tears such as kneeling near 90° of flexion.

2. A kneel-assist device should be developed, by manufacturers, to reduce the possible negative consequences of kneeling near full flexion, which includes high flexion moments and increased levels of internal tibial rotation. Additionally, workers should focus on keeping their ankles straight when kneeling near full flexion, to reduce tibial rotations and thereby reduce loading to the meniscus.

3. Squatting should be used sparingly and is not recommended.

4. When performing static tasks, avoid kneeling on one knee. Instead, kneeling near full flexion or kneeling near 90° of knee flexion should be adopted.

5. Kneepads must be redesigned such that they signifiantly reduce the pressure at the patella and, especially, at the PTT region. Manufacturers should obtain feedback from mine workers regarding commonly reported problems such as strap discomfort and coal particles trapped in the kneepad. Furthermore, manufacturers may want to consider a comprehensive approach whereby kneepads and other kneel-assist devices (e.g., foam wedge behind worn between the calf and thigh) are designed as one system.

Several limitations to this study should be acknowledged. One limitation is the subjects were not experienced mine workers. Also, only two seam heights and three kneepad conditions were tested in a limited number of postures for this study. Furthermore, pressure data could not be determined for the dynamic postures (i.e., twisting at the waist) due to data collection limitations. Despite these limitations, information received through in-mine observations and feedback from mine workers indicated that the conditions tested would represent the majority of tasks performed by low-seam mine workers. A further limitation was the size of the gaps between sensing units of the sensor resulting from the need for the sensor to be curved in design. The largest gap areas were in the central portion of the curved section of the sensor pad, or where the superior region of the patella resided. Because the inferior border of the patella was the only

portion experiencing loading, the impact of these larger gaps is likely minimal. However, hot spots (i.e. areas of high pressure) may have been missed in the PTT region. Therefore, the results of this study likely underestimated the peak and mean pressures in this region. Knee kinematics were determined in 3-D; however, due to the limitations of the pressure sensor, thigh-calf and heel-gluteus contact forces were obtained in
2-D. These contact forces were measured normal to the tibia and thus the medial-lateral and superior-inferior shear forces caused by tissue deformation were neglected, thereby affecting the accuracy of the estimated shear forces, internal rotation, and varus moments. Due to the application of kneepads and the custom knee-pressure sensor, it was not possible to track the positions of the knee joint center during testing. Instead, an anatomical coordinate system was constructed to estimate the positions of these markers from the testing marker set worn by subjects during testing. This introduced mean errors of less than 4% for the kneeling postures and less than 9% for the squatting postures. However, due to the complexity of high-flexion kneeling, using markers to approximate the knee joint center for kneeling near full flexion and squatting may not have been feasible even without kneepads as femoral rollback causes an internal rotation of the tibia. When this occurs, the knee joint center may only be approximated by imaging techniques which were not employed in this study. In effect, some inaccuracies may be present with the tibial moments, as femoral rollback and knee joint center errors affect the moment arm calculations. However, the 3-D model used in this study may minimize these errors by accounting for tibial rotations. Finally, back muscles were not investigated in this study; however, the authors do not consider this a strong limitation as previous research has demonstrated that the knee is injured 1.7 times more often than the back in low-seam coal mines. In addition, extensive literature already exists regarding the lifting capacity of workers when in kneeling and squatting postures in regard to back stressors [Ayoub, et al. 1985a, Ayoub, et al. 1985b, Gallagher and Hamrick 1992, Gallagher, et al. 2009, Gibbons 1989].

Conclusions

Researchers should continue to investigate the tasks performed by low-seam mine workers and the postures associated with these tasks. The data obtained in future studies, coupled with the data reported in the current study, may then be used to develop an appropriate and practical postural rotation strategy for the low-seam mining occupation. Squatting and kneeling on one knee may be included in a postural rotation strategy upon occasion, but long bouts of exposure to these postures should be avoided. Some mining tasks may dictate the posture a mine worker is capable of adopting, making it difficult, or impossible, to use a postural rotation strategy. In these cases, mine workers should take occasional breaks during a shift to move their knees through the entire range of motion (i.e., flush the knee) [Moore, et al. 2008].

Manufacturers should consider a novel approach to the design of personal protective equipment for individuals who kneel for extended periods of time, including low-seam mine workers. A kneepad design that transfers pressure at the knee to other parts of the body that have larger surface areas (e.g., the shin) should be considered. Furthermore, manufacturers should consider the benefits of providing kneel-assist devices in addition to kneepads (e.g., foam wedge worn between calf and thigh). Kneeling near full flexion resulted in the least amount of muscle activity. Interviews with mine workers indicated that this posture is likely the most predominant posture used which may be a direct result of lower energy expenditures in this posture due to reduced muscular demands. Therefore, manufacturers should consider the development of devices that address knee loading during this specific posture. Such devices may include a cushion in the shape of a wedge worn at the ankle such that the applied flexion moment at the knee is reduced when kneeling near full flexion. Additionally, these devices may include a bracing feature to reduce the internal rotation of the tibia during kneeling, which may alleviate tension in the ligaments and reduce the loading on the medial compartment of the tibia.

Acknowledgments

The authors would like to acknowledge the contributions of Mary Ellen Nelson and Albert Cook for their assistance in data collection.

References

Ayoub MM, Smith JL, Selan JL, Fernandez JE [1985a]. Manual materials handling in unusual positions—Phase I. Final report prepared for the University of Dayton Research Institute.

Ayoub MM, Smith JL, Selan JL, Chen HC, Fernandez JE, Lee YH, Kim HK [1985b]. Manual materials handling in unusual positions—Phase II. Final report prepared for the University of Dayton Research Institute.

Baker P, Coggon D, Reading I, Barrett D, McLaren M [2002]. Sports injury, occupational physical activity, joint laxity, and meniscal damage. J Rheumatol 29:557–563.

Baker P, Reading I, Cooper C, Coggon D [2003]. Knee disorders in the general population and their relation to occupation. Occup Environ Med 60:794–797.

Bell A, Brand RA, Pedersen DR [1990]. A comparison of the accuracy of several hip center location prediction methods. J Biomech 23:617–621.

Bender R and Lange S [2001]. Adjusting for multiple testing—when and how? J Clin Epidemiol 54:343–349.

Coggon D, Croft P, Kellingray S, Barrett D, McLaren M, Cooper C [2000]. Occupational physical activities and osteoarthritis of the knee. Arthritis & Rheum 43:1443–1449.

Cooper C, McAlindon T, Coggan D, Egger P, Dieppe P [1994]. Occupational activity and osteoarthritis of the knee. Ann Rheum Dis 53:90–93.

Ericson MO, Nisell R, Arborelius UP, Ekholm J [1985]. Muscular activity during ergometer cycling. Scand J Rehabil Med 17(2):53–61.

Felson DT, Hannan MT, Naimark A, Berkeley J, Gordon G, Wilson PW, Anderson J [1991]. Occupational physical demands, knee bending, and knee osteoarthritis: results from the Framingham Study. J Rheumatol 18:1587–1592.

Gallagher S and Hamrick CA [1992]. Acceptable workloads for three common mining materials. Ergonomics 35(9):1013–1031.

Gallagher S, Moore S, Dempsey P [2009]. An analysis of injury claims from low-seam coal mines. J Safety Research 40:233–237.

Gibbons LE [1989]. Summary of ergonomics research for the crew chief model development: interim report for period February 1984 to December 1989. Armstrong Aerospace Medical Research Laboratory, report no. AAMRL-TR-90-038. Wright-Patterson Air Force Base, Dayton, Ohio, 390 pp.

Hinrichs RN [1990]. Adjustments to the segment center of mass proportions of Clauser et al. (1969). J Biomech 23(9):949–951.

Kirtley C [2006]. Clinical gait analysis: theory and practice. Churchill Livingstone: Elsevier, pp. 57–61.

Mayton AG, Porter WL, Pollard J, Moore SM [2010]. Pressure on the knee while performing a lateral lift from kneeling postures. Annual American Society of Biomechanics Meeting. Providence, Rhode Island. August 19–21.

McMillan G and Nichols L [2005]. Osteoarthritis and meniscus disorders of the knee as occupational diseases of miners. Occup Environ Med 62(8):567–575.

Moore SM, Bauer ER, Steiner LJ [2008]. Prevalence and cost of cumulative injuries over two decades of technological advances: a look at underground coal mining in the U.S. Min Eng 60(1):46–50.

MSHA [2009]. Accident, illness and injury and employment self-extracting files (part 50 data). Denver, CO: U.S. Department of Labor, Mine Safety and Health Administration, Office of Injury and Employment Information. [http://www.msha.gov/STATS/PART50/p50y2k/p50y2k.HTM]. Date accessed: August 2009.

Nagura T, Dyrby CO, Alexander EJ, Andriacchi TP [2002]. Mechanical loads at the knee joint during deep flexion. J Orthop Res 20:881–886.

NIOSH [1992]. Selected topics in surface electromyography for use in the occupational setting: expert perspectives. By Soderber GL, ed. Cincinnati, OH: U.S. Department of Health and Human Services, Centers for Disease Control, National Institute for Occupational Safety and Health, DHHS (NIOSH) Publication No. 91–100 179.

Perry J, Antonelli D, Ford W [1975]. Analysis of knee-joint forces during flexed-knee stance. J Bone Joint Surgery 57:961–967.

Roantree WB [1957]. A review of 102 cases of beat conditions of the knee. Brit J Industr Med 14:253–257.

Sharrard WJW [1964]. Pressure effects on the knee in kneeling miners. Royal College of Surgeons of England: pp. 309–324.

Sharrard WJ [1963]. Aetiology and pathology of beat knee. Brit J Industr Med 20:24–31.

Sharrard WJ and Liddell FD [1962]. Injuries to the semilunar cartilages of the knee in miners. Br J Ind Med 19:195–202.

Tanaka S, Halperin WE, Smith AB, Lee ST, Luggen ME, Hess EV [1985]. Skin effects of occupational kneeling. Am J Ind Med 8:341–349.

Watkins JT, Hunt TA, Fernandez RH, Edmonds OP [1958]. A clinical study of beat knee. Br J Ind Med *15*(2):105–109.

Winter DA [1990]. Biomechanics and motor control of human movement. New York: John Wiley & Sons, Inc.: 176–199.

Appendix A

For each segment, an anatomical and measured coordinate system was created from the motion capture data. The anatomical system was created from the anatomical standing T-pose and allowed the location of anatomical landmarks to be linked to the global reference frame. It was also used to determine the location of the ankle joint center (AJC), knee joint center (KJC), and hip joint center (HJC) as well as the location of the lower leg center of mass. A measured coordinate system was created from the anatomical standing T-pose as well as from each static trial, and was then used to link the testing markers to the locations of the markers that were removed.

The anatomical coordinate system of the thigh (ATCS) was created using the left and right anterior superior iliac spine (L.ASIS & R.ASIS), knee, and thigh markers.

$$r_1 = \frac{knee\ lateral - knee\ medial}{|knee\ lateral - knee\ medial|} \quad x-axis$$

$$r_2 = \frac{HJC - KJC}{|HJC - KJC|}$$

$$r_3 = r_2 \times r_1 \quad y-axis$$

$$r_4 = r_1 \times r_3 \quad z-axis$$

The KJC was assumed to be midway between the medial and lateral epicondyles of the femur, measured by the medial and lateral knee markers. The location of the HJC was approximated using regression equations proposed by Bell et al. [1990] and adapted to fit the global reference frame of the laboratory. [Bell 1990] (Figure 22)

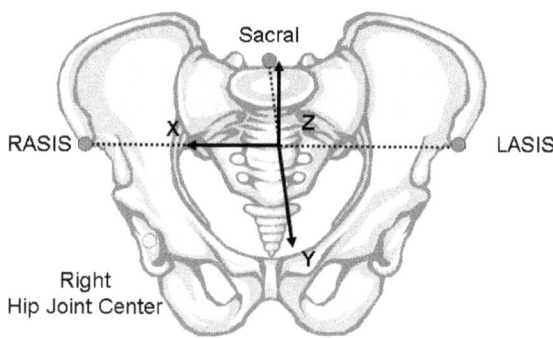

Figure 22: Pelvis coordinate system highlighting the location of the right hip joint center

$$Origin = \frac{LASIS + RASIS}{2}$$

$$PW = |RASIS_x - LASIS_x|$$

$$HJC = \begin{bmatrix} Origin + .36*PW \\ Origin - .19*PW \\ Origin - .3*PW \end{bmatrix}$$

The transformation matrix from the global reference frame to the ATCS (T_{TGA}) was created from the unit direction vectors of the ATCS.

$$T_{TGA} = \begin{bmatrix} \begin{bmatrix} 1 \\ HJC \end{bmatrix} & \begin{bmatrix} 0 \\ r_1 \end{bmatrix} & \begin{bmatrix} 0 \\ r_3 \end{bmatrix} & \begin{bmatrix} 0 \\ r_4 \end{bmatrix} \end{bmatrix}$$

The anatomical coordinate system of the shank (ASCS) was determined using the knee, ankle, and shank markers. The AJC was assumed to be midway between the medial and lateral malleoli, measured by the medial and lateral ankle markers. The transformation matrix from the GCS to the ASCS (T_{SGA}) was created from the unit direction vectors of the ASCS.

$$r_1 = \frac{knee\ lateral - knee\ medial}{|knee\ lateral - knee\ medial|}$$

$$r_2 = \frac{KJC - AJC}{|KJC - AJC|} \quad\quad z-axis$$

$$r_3 = r_1 \times r_2 \quad\quad y-axis$$

$$r_4 = r_3 \times r_1 \quad\quad x-axis$$

$$T_{SGA} = \begin{bmatrix} \begin{bmatrix} 1 \\ KJC \end{bmatrix} & \begin{bmatrix} 0 \\ r_4 \end{bmatrix} & \begin{bmatrix} 0 \\ r_3 \end{bmatrix} & \begin{bmatrix} 0 \\ r_1 \end{bmatrix} \end{bmatrix}$$

The ATCS and ASCS were oriented such that when standing the systems aligned with the GCS and the positive x-axis was in the lateral direction of the right leg, the positive z-axis was in the proximal direction, and the positive y-axis was in the anterior direction. (Figure 23)

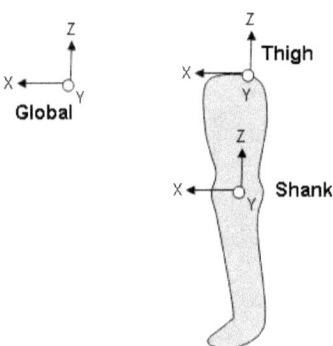

Figure 23: Orientation of the ATCS and ASCS

A measured coordinate system (MCS) was created for the thigh and the shank using the marker clusters on the segments. The measured coordinate system of the thigh (MTCS) was created from the thigh, thigh front, and thigh rear markers. The transformation matrix from the GCS to the MTCS (T_{TGM}) was also created with the right thigh front marker as its origin.

$$r_1 = \frac{thigh - thigh\ front}{|thigh - thigh\ front|} \qquad z-axis$$

$$r_2 = \frac{thigh\ front - thigh\ rear}{|thigh\ front - thigh\ rear|}$$

$$r_3 = r_2 \times r_1 \qquad x-axis$$

$$r_4 = r_1 \times r_3 \qquad y-axis$$

$$T_{TGM} = \begin{bmatrix} 1 & 0 & 0 & 0 \\ thigh\ front & r_3 & r_4 & r_1 \end{bmatrix}$$

The measured coordinate system of the shank (MSCS) was created from the shank, shank front, and shank rear markers. The transformation matrix from the MSCS to the GCS (T_{SGM}) was created with the right shank front marker as its origin.

$$r_1 = \frac{shank - shank\ front}{|shank - shank\ front|} \qquad z-axis$$

$$r_2 = \frac{shank\ front - shank\ rear}{|shank\ front - shank\ rear|}$$

$$r_3 = r_2 \times r_1 \qquad x-axis$$

$$r_4 = r_1 \times r_3 \qquad y-axis$$

$$T_{SGM} = \begin{bmatrix} \begin{bmatrix} 1 \\ shank\ front \end{bmatrix} & \begin{bmatrix} 0 \\ r_3 \end{bmatrix} & \begin{bmatrix} 0 \\ r_4 \end{bmatrix} & \begin{bmatrix} 0 \\ r_1 \end{bmatrix} \end{bmatrix}$$

To determine measured marker locations in the ATCS, T_{TMA} was created; for the ASCS, T_{SMA} was created.

$$T_{TMA} = [T_{TGM}]^{-1}[T_{TGA}]$$

$$T_{SMA} = [T_{SGM}]^{-1}[T_{SGA}]$$

Appendix B

A_{ij} : Matrix of the cells that make up the patella region (see Figure 4)
B_{ij} : Matrix of the cells that make up the PTT region (see Figure 4)
i : sample number
j : cell number
n : total number of samples
m : total number of sensing units in B (PTT)
q : total number of sensing units in A (patella)

Mean pressure ratio: the average over time of the ratio of the sum of a region over the sum over both regions.

$$\frac{\sum_{i=1}^{n} \frac{\sum_{j=1}^{m} B_{ij}}{\sum_{j=1}^{q} A_{ij} + \sum_{j=1}^{m} B_{ij}}}{n}$$

Pressure mean of mean: the average of a region averaged over time

$$\frac{\sum_{i=1}^{n} \frac{\sum_{j=1}^{m} B_{ij}}{m}}{n}$$

Pressure mean of max: the maximum of a region averaged over time

$$\frac{\sum_{i=1}^{n} \max(B_i)}{n}$$

Pressure mean of variance: the variance of a region averaged over time

$$\frac{\sum_{i=1}^{n} (stdev(B_i))^2}{n}$$

NOTE: Equations for the patellar tendon and tibial tubercle region are shown. Patella equations are similar.

www.ingramcontent.com/pod-product-compliance
Lightning Source LLC
Chambersburg PA
CBHW081905170526
45167CB00007B/3153